Gas Biology Research in Clinical Practice

Gas Biology Research in Clinical Practice

Editors

Toshikazu Yoshikawa Kyoto
Yuji Naito Kyoto

32 figures and 11 tables, 2011

Basel · Freiburg · Paris · London · New York · Bangalore · Bangkok · Shanghai · Singapore · Tokyo · Sydney

Toshikazu Yoshikawa
Molecular Gastroenterology and Hepatology
Graduate School of Medical Science
Kyoto Prefectural University of Medicine
465 Kajiicho Hirokoji Kawaramachi-dori,
Kamigyo-ku
Kyoto 602-8566
Japan

Yuji Naito
Molecular Gastroenterology and Hepatology
Graduate School of Medical Science
Kyoto Prefectural University of Medicine
465 Kajiicho Hirokoji Kawaramachi-dori,
Kamigyo-ku
Kyoto 602-8566
Japan

Library of Congress Cataloging-in-Publication Data

Gas biology research in clinical practice / editors, Toshikazu Yoshikawa, Yuji Naito.
p. ; cm.
Includes bibliographical references and indexes.
ISBN 978-3-8055-9664-0 (hard cover : alk. paper) -- ISBN 978-3-8055-9665-7 (e-ISBN)
1. Blood gases. 2. Gases--Physiological effect. 3. Nitric oxide--Physiological effect. 4. Carbon monoxide--Physiological effect. 5. Hydrogen sulfide--Physiological effect. I. Yoshikawa, Toshikazu, 1947- editor. II. Naito, Yuji, editor.
[DNLM: 1. Gases--pharmacology. 2. Gases--diagnostic use. 3. Gases--metabolism. 4. Gases--therapeutic use. QV 310]
QP99.3.G3G375 2011
572'.54--dc22

2010048614

Bibliographic Indices. This publication is listed in bibliographic services.

www.karger.com
Printed in Switzerland on acid-free and non-aging paper (ISO 9706) by Reinhardt Druck, Basel
ISBN 978–3–8055–9664–0
e-ISBN 978–3–8055–9665–7

Contents

VII Preface
Yoshikawa, T.; Naito, Y. (Kyoto)

General

1 Roles of Stress-Inducible Carbon Monoxide in the Regulation of Liver Function
Suematsu, M.; Kajimura, M.; Kabe, Y. (Tokyo)

6 Intraluminal Gas and Gastrointestinal Diseases
Urita, Y.; Watanabe, T.; Takemoto, I.; Sasaki, Y.; Maeda, T.; Matsumoto, M.; Honda, Y.; Shimada, N.; Nakajima, H.; Sugimoto, M. (Tokyo)

15 Therapeutic Medical Gas
Nakao, A. (Pittsburgh, Pa.)

Gas and Medical Application: I. CO

24 Analysis of Breath CO and Application to Hemodynamic Monitoring
Sawano, M. (Saitama)

35 CO and Its Application to Gastrointestinal Disease
Naito, Y.; Uchiyama, K.; Takagi, T.; Yoshikawa, T. (Kyoto)

Gas and Medical Application: II. NO

43 Clinical Application of Inhaled Nitric Oxide
Maruyama, K.; Zhang, E.; Maruyama, J. (Mie)

56 Inhaled Nitric Oxide and Clinical Application
Shime, N.; Inoue, M.; Hashimoto, S. (Kyoto)

Gas and Medical Application: III. H_2S

65 Hydrogen Sulfide in the Gastrointestinal Tract: Friend or Foe?
Ock, C.Y.; Han, S.J.; Choi, K.-S.; Kim, E.-H.; Kim, J.H.; Hahm, K.-B. (Incheon)

73 Role of Hydrogen Sulfide in Colitis
Takagi, T.; Naito, Y.; Hirata, I.; Yoshikawa, T. (Kyoto)

81 HCO_3^- Stimulatory Action of Hydrogen Sulfide in Rat Duodenum
Takeuchi, K.; Ise, F.; Koyama, M.; Dogishi, K.; Yasuda, M.; Hayashi, S. (Kyoto)

Gas and Medical Application: IV. H_2

91 **Hydrogen and Medical Application**
Nakao, A. (Pittsburgh, Pa.)

Gas and Medical Application: V. ^{13}C

100 **^{13}C-Breath Test for Studying Physiology and Pathophysiology by Using Experimental Animals**
Uchida, M. (Kanagawa)

112 **Carbon-13 and Its Clinical Application**
Tando, Y.; Matsumoto, A.; Matsuhashi, Y.; Tanaka, H.; Yanagimachi, M.; Nakamura, T. (Aomori)

Gas and Medical Application: VI. Others

119 **Acetone Response during Graded and Prolonged Exercise**
Sasaki, H.; Ishikawa, S.; Ueda, H.; Kimura, Y. (Osaka)

125 **Findings of Skin Gases and Their Possibilities in Healthcare Monitoring**
Tsuda, T.; Ohkuwa, T.; Itoh, H. (Nagoya)

133 **Phytoncide – Its Properties and Applications in Practical Use**
Nomura, M. (Hiroshima)

144 **Author Index**

145 **Subject Index**

Preface

Since 2004 we have held annually in Kyoto, Japan, a medical conference on heme oxygenase (HO), titled the 'HO Research Forum'. To date, more than 80 papers have been presented at these conferences, and our knowledge of HO and carbon monoxide, one of the by-products by HO, has been greatly expanded. Recently, evidence has accumulated suggesting that carbon monoxide plays an important role in many physiological and pathological conditions. Thus, in the last decade there has been an extraordinarily rapid growth in our knowledge of gaseous molecules such as molecular oxygen, nitric oxide, hydrogen sulfide and carbon monoxide. These gaseous molecules have been shown to play important roles in signal transduction in biological systems.

This book is an attempt to highlight some of the impressive recent advances in gas biology. It was initiated by Dr. Hideo Ueda, but sadly illness prevented him from completing the project. As the subject matter of this book is of importance for most basic and clinical researchers involved in gas biology, we decided to take it on. We were fortunate to receive excellent manuscripts from authors in the field, and readers will find they contain many new insights into leading-edge research on gas biology.

We would like to thank the many authors and colleagues who have contributed to the success of this publication. Special thanks also go to Dr. Tomohisa Takagi for his assistance. Finally, we thank Karger Publishers for their cooperation and encouragement throughout the publication process.

Toshikazu Yoshikawa
Yuji Naito

Yoshikawa T, Naito Y (eds): Gas Biology Research in Clinical Practice.
Basel, Karger, 2011, pp 1–5

Roles of Stress-Inducible Carbon Monoxide in the Regulation of Liver Function

Makoto Suematsu · Mayumi Kajimura · Yasuaki Kabe

Department of Biochemistry, School of Medicine, Keio University, Japan Science and Technology Agency, ERATO Suematsu Gas Biology Project, Tokyo, Japan

Abstract

Carbon monoxide (CO) is a gaseous product generated by heme oxygenase (HO). As the liver is a gigantic resource of CO derived from heme degradation in vivo, constitutive and inducible CO has been suggested to regulate porto-sinusoidal vascular as well as biliary function. Previous studies revealed that CO generated by HO modulates function of different heme proteins or enzymes through binding to their prosthetic ferrous heme to regulate the biological function of the hepatobiliary systems; the proteins targeted by the gas involve soluble guanylate cyclase, cytochromes P450 and cystathionine β-synthase. Because of the heterogeneous distribution of these enzymes in the liver tissue, CO regulates liver functions through multiple mechanisms that protect the tissue against varied noxious stimuli. The current article overviews the intriguing regulatory mechanisms operated by CO and their medical implications.

CO Derived from HO-1 in Macrophages

CO is a nonradical gaseous mediator generated by a mono-oxygenase reaction through heme oxygenase (HO). It was first shown in the early 1970s that both non-parenchymal cells and hepatocytes enable the degradation of heme to bile pigments. Kupffer cells play a role in the removal and degradation of senescent erythrocytes, while hepatocytes catabolize free heme derived from hemoglobin or cytochromes P450 [1]. Distribution of HO-1 and HO-2 in the liver was previously examined by our laboratory in rats and humans, indicating that the two isozymes have distinct topographic patterns: HO-1, the inducible form, is prominent in Kupffer cells, while the constitutive HO-2 is mostly abundant in hepatocytes [2, 3]. Our studies using a specific monoclonal antibody against HO-1 that blocks the isozyme activity revealed that approximately 70% of the whole HO activity in this tissue is derived from HO-2 [3]; considering the percentage of the cell population of Kupffer cells in the liver, these cells possess huge amounts of HO activity as compared with hepatocytes.

Not only in the liver but also in other organs, resident macrophages constitute a major site of the HO-1 distribution. To better understand the tissue iron overload and anemia previously reported in a human patient and mice that lack HO-1, recent studies examined iron distribution and pathology in HO-1(–/–) mice [4], indicating that resident splenic and liver macrophages were mostly absent in HO-1(–/–) mice. This study suggested that erythrophagocytosis causes death of HO-1(–/–) macrophages in vivo and release of nonmetabolized heme likely caused tissue inflammation. In the spleen, initial splenic enlargement progressed to red pulp fibrosis, atrophy and functional hyposplenism in older mice, recapitulating the asplenia of an HO-1-deficient patient which was previously reported in Japan [5]. During constitutive erythrophagocytosis, HO-1-deficient macrophages including Kupffer cells appeared to exhibit cell death by themselves. This study also postulated that failure of tissue macrophages to remove senescent erythrocytes led to intravascular hemolysis and increased expression of hemopexin and haptoglobin, the heme and hemoglobin scavenger proteins, in the liver. Lack of macrophages expressing the Hp receptor, CD163, diminished the ability of Hp to neutralize circulating Hb, and iron overload occurred in kidney proximal tubules, which were able to catabolize heme with HO-2. Thus, in HO-1(–/–) mammals, the reduced function and viability of erythrophagocytosing macrophages appear to be the main causes of tissue damage and iron redistribution.

CO Derived from HO-2 Targets Hepatic Stellate Cells to Relax Sinusoids

Many years ago, we first showed that CO serves as a vasorelaxing factor that reduces sinusoidal tone and is necessary for maintenance of the microvascular perfusion [6, 7]. This is based on the fact that zinc protoporphyrin IX, a potent inhibitor of HO, abolished CO generation and simultaneously increased the baseline vascular resistance in the perfused liver. The role of endogenous CO in sinusoidal relaxation was also shown in experiments using free oxyhemoglobin which captures both NO and CO, and methemoglobin that is able to capture NO but not CO [2]: only oxyhemoglobin was able to reproduce the vasoconstrictive effect of the HO inhibitor. Although the gas appears to diffuse freely across the cell membrane, our results collected from the liver did not support such a notion: when encapsulated with liposome, oxyhemoglobin lost its ability to constrict sinusoids because of limited accessibility to the space of Disse across the endothelial fenestration, suggesting that the locus of the gas action is extrasinusoidal, perhaps involving hepatic stellate (Ito) cells. Ito cells constitute the most abundant resource of soluble guanylate cyclase in the liver that serves as a receptor for CO-responsive modest upregulation of cyclic GMP through the gas binding to the prosthetic ferrous heme of proteins.

Although a lot of evidence suggests that CO generated from HO plays a protective role against organ dysfunction, few studies have provided evidence for quantitative information of endogenous CO generation: according to previous data, local concentrations of CO in the liver under physiologic conditions appeared to be around 1 μmol/l. Under pathologic conditions including ischemia-reperfusion, acetaminophen-induced liver injury [8], endotoxemia [9] or excessive heme overloading [10], the CO concentrations reached 2–4 μmol/l. Still unclear is the extent to which NO could interact or compete with CO in vivo, mainly because of the lack of quantitative information on functionally intact NO in situ. Although sinusoidal endothelium constitutes a major cellular component of NO synthase expression, NO released from the enzyme may be entrapped by circulating erythrocytes or cancelled by superoxide anion spontaneously released from Kupffer cells [11].

Since the potency of NO to activate the cyclase is extremely greater than that of CO [12], CO appears to be able to activate soluble guanylate cyclase in vivo, but only when local NO concentrations are low. Actually, when the liver is exposed to endotoxemia to induce inducible NO synthase [9], amounts of cyclic GMP are dictated by NO but not CO. Since cyclic nucleotides upregulate the transcriptional expression of HO-1, the endotoxemic liver overexpresses HO-1; this response downregulates inducible NO synthase through degradation of the prostethic heme for this enzyme to suppress NO, and subsequently upregulates CO to maintain sinusoidal relaxation for the blood supply. Under these circumstances, CO relaxes sinusoids through cyclic GMP-independent mechanisms [9]. In this relaxation mechanism, the ability of CO to inhibit cytochrome P450 epoxygenases was suggested to be involved: in other other words, the stress-inducible CO is necessary to maintain sinusoidal blood flow and subsequently to guarantee bile output under disease conditions.

CBS as a CO Sensor Mined by Metabolomic Analyses

Because of the nature of the gas to bind to the metal-centered prosthetic groups of macromolecules, it is not unreasonable to hypothesize that biological gases bind to enzymes in metabolic systems that constitute a major class of such proteins. Based on this assumption, we have recently applied metabolomic analyses assisted by capillary electrophoresis combined with mass spectrometry (CE-MS) in order to mine footprints of the gases of interest on alterations in small molecular metabolites in the metabolic systems [13, 14]. In recent studies, we examined the effects of CO on the metabolic systems using several different experimental models where the gas was upregulated significantly. Results suggested that CO has the ability to inhibit the transsulfuration pathway that is rate-limited by cystathionine β-synthase (CBS), a heme-containing enzyme [14].

Based on these data, we examined the roles of the enzyme for a CO-specific receptor candidate in vivo. Several lines of biochemical evidence support the concept that CBS acts as a CO sensor. First, studies using recombinant CBS have shown that CO inhibits CBS with a K_i value of approximately 5 μmol/l [15], being comparable to the concentration occurring in the liver. Second, murine hepatocytes express both

CO-producing HO and H_2S-producing CBS with additional HO-1 induced in both hepatocytes and Kupffer cells under stress conditions. The close proximity of the enzyme distributions taken together with measured CO concentrations in the models, and the kinetics of CBS activity led us to hypothesize that CBS is acting as a CO sensor in vivo. In vivo pulse-chase analyses suggest that CBS is the enzyme that actually modulates the metabolic flux of this pathway, as judged so far from the results collected by ^{15}N-methionine flux analyses using CE-MS [14]. To note is that CBS is the enzyme that generates H_2S. CO-overproducing livers showed a decrease in labile H_2S amount, whereas the livers of heterozygous CBS knockout mice did not show any notable decrease in H_2S in response to stress-inducible levels of CO, suggesting that the gas inhibits the activity of CBS in vivo. Such a stress-inducible suppression of H_2S in the liver stimulates HCO_3^--dependent choleresis that helps the solubility of organic anions in bile [14]. Mechanisms by which H_2S modulates biliary excretion might involve glibenclamide-sensitive Na^+-K^+-$2Cl^-$ channels in the biliary system, although whether the gas might directly bind to the channel remains unknown.

We believe that the gas involving CO has multiple modes of biological actions through varied specific receptors. CBS appears to be one of such multiple receptors that regulate organ functions. In order to mine novel receptors that have not yet been identified, further investigation with different technological approaches should obviously be necessary.

Acknowledgments

This work was supported by JST, ERATO, Suematsu Gas Biology Project, Tokyo 160–8582. Establishment of metabolomic analysis was supported by a Global COE Project for Human Metabolomics Systems Biology as well as by Research and Development of the Next-Generation Integrated Simulation of Living Matter, a part of the Development and Use of the Next-Generation Supercomputer Project of MEXT.

References

1 Tenhunen R, Marver HS, Schmid R: The enzymatic conversion of heme to bilirubin by microsomal heme oxygenase. Proc Natl Acad Sci USA 1968;61: 748–755.

2 Goda N, Suzuki K, Naito M, Takeoka S, Tsuchida E, Ishimura Y, Tamatani T, Suematsu M: Distribution of heme oxygenase isoforms in rat liver: topographic basis for carbon monoxide-mediated microvascular relaxation. J Clin Invest 1998;101:604–612.

3 Makino N, Suematsu M, Sugiura Y, Morikawa H, Shiomi S, Goda N, Sano T, Nimura Y, Sugimachi K, Ishimura Y: Altered expression of heme oxygenase-1 in the livers of patients with portal hypertensive diseases. Hepatology 2001;33:32–42.

4 Kovtunovych G, Eckhaus MA, Ghosh MC, Ollivierre-Wilson H, Rouault TA: Dysfunction of the heme recycling system in heme oxygenase-1 deficient mice: effects on macrophage viability and tissue iron distribution. Blood 2010 Oct 1. [Epub ahead of print]

5 Yachie A, Niida Y, Wada T, Igarashi N, Kaneda H, Toma T, Ohta K, Kasahara Y, Koizumi S: Oxidative stress causes enhanced endothelial cell injury in human heme oxygenase-1 deficiency. J Clin Invest 1999;103:129–135.

6 Suematsu M, Kashiwagi S, Sano T, Goda N, Shinoda Y, Ishimura Y: Carbon monoxide as an endogenous modulator of hepatic vascular perfusion. Biochem Biophys Res Commun 1994;205:1333–1337.
7 Suematsu M, Goda N, Sano T, Kashiwagi S, Egawa T, Shinoda Y, Ishimura Y: Carbon monoxide: an endogenous modulator of sinusoidal tone in the perfused rat liver. J Clin Invest 1995;96:2431–2437.
8 Mori M, Suematsu M, Kyokane T, Sano T, Suzuki H, Yamaguchi T, Ishimura Y, Ishii H: Carbon monoxide-mediated alterations in paracellular permeability and vesicular transport in acetaminophen-treated perfused rat liver. Hepatology 1999;30:160–168.
9 Kyokane T, Norimizu S, Taniai H, Yamaguchi T, Takeoka S, Tsuchida E, Naito M, Nimura Y, Ishimura Y, Suematsu M: Carbon monoxide from heme catabolism protects against hepatobiliary dysfunction in endotoxin-treated rat liver. Gastroenterology 2001;120:1227–1240.
10 Wakabayashi Y, Takamiya R, Mizuki A, Kyokane T, Goda N, Yamaguchi T, Takeoka S, Suematsu M, Ishimura Y: Carbon monoxide overproduced by heme oxygenase-1 causes a reduction of vascular resistance in perfused rat liver. Am J Physiol Gastrointest Liver Physiol 1999;277:G1088–G1096.
11 Bautista AP, Spitzer JJ: Inhibition of nitric oxide formation in vivo enhances superoxide release by the perfused liver. Am J Physiol 1994;266:G783-G788.
12 Kharitonov VG, Sharma VS, Pilz RB, Magde D, Koesling D: Basis of guanylate cyclase activation by carbon monoxide. Proc Natl Acad Sci USA 1995; 92:2568–2571.
13 Soga T, Baran R, Suematsu M, Ueno Y, Ikeda S, Sakurakawa T, Kakazu Y, Ishikawa T, Robert M, Nishioka T, Soga T: Differential metabolomics reveals ophthalmic acid as an oxidative stress biomarker indicating hepatic glutathione consumption. J Biol Chem 2006;281:16768–16776.
14 Shintani T, Iwabuchi T, Soga T, Kato Y, Yamamoto T, Takano N, Hishiki T, Ueno Y, Ikeda S, Sakuragawa T, Ishikawa K, Goda N, Kitagawa Y, Kajimura M, Matsumoto K, Suematsu M: Cystathionine β-synthase as a carbon monoxide-sensitive regulator of bile excretion. Hepatology 2009;49:141–150.
15 Taoka S, Banerjee R: Characterization of NO binding to human cystathionine β-synthase: possible implications of the effects of CO and NO binding to the human enzyme. J Inorg Biochem 2001;87:245–251.

Makoto Suematsu, MD
Department of Biochemistry, School of Medicine, Keio University
35 Shinanomachi, Shinjuku-ku
Tokyo 160-8582 (Japan)
Tel. +81 3 5363 3753, Fax +81 3 5363 3466, E-Mail msuem@sc.itc.keio.ac.jp

Yoshikawa T, Naito Y (eds): Gas Biology Research in Clinical Practice.
Basel, Karger, 2011, pp 6–14

Intraluminal Gas and Gastrointestinal Diseases

Yoshihisa Urita · Toshiyasu Watanabe · Ikutaka Takemoto · Yosuke Sasaki · Tadashi Maeda · Manabu Matsumoto · Yoshiko Honda · Nagato Shimada · Hitoshi Nakajima · Motonobu Sugimoto

Department of General Medicine and Emergency Care, Toho University, School of Medicine, Tokyo, Japan

Abstract

Gases are also produced while passing through the gastrointestinal tract, possibly resulting in abdominal symptoms although gas is continuously removed by eructation, anal evacuation, absorption through the intestinal mucosa, and bacterial consumption. It seems that 2–8 liters of air may be swallowed and eructated per day. Approximately 5–10 liters of CO_2 seems to be produced and released into the lumen by chemical reactions in the upper gut. The pCO_2 rises dramatically in the duodenum, resulting in diffusion of CO_2 from lumen to blood, whereas CO_2 in swallowed air diffuses from the blood into the stomach. The pN_2 of swallowed air is slightly higher than that of venous blood in the stomach, possibly resulting in very slow absorption. Since pH_2 and pCH_4 are always higher in the lumen than in the blood, gases are constantly diffusing from the lumen to the blood. The calculated amount of H_2 produced in the colon is 2.7–27 liters per day because H_2 gas is produced at a rate of 4 liters for every 12.5 g of undigested carbohydrate. Since CH_4 production occurs primarily in the left colon whereas H_2 is produced primarily in the right colon, H_2 produced in the left colon may be rapidly converted to CH_4, possibly resulting in reduced flatus. Greater amount of gases pass through the digestive tract per day than liquid and solid contents, suggesting that impaired gas movement might be more closely associated with abdominal symptoms compared to liquid movement.

The overall function of the digestive system is to transfer the nutrients in food from the external environment to the internal environment. During the course of the day, an adult usually consumes about 800 g of food and 1.5 liters of water and most of the nutrients and water are normally absorbed in the small intestine. Only approximately 150 g are eliminated from the body as feces. Secretion into the tract may amount to 7–8 liters of fluid, resulting in a total of 9–10 liters of fluid passing through the tract per day. Thus, a great amount of gases enter or leave the gastrointestinal tract with both solid and liquid substances.

On the other hand, although the intestines of normal subjects usually contain less than 200 ml of gas regardless of taking meals, the gastric bubble present in approximately 70% of normal chest and abdominal radiograph results from the 2–3 ml of air conveyed to the stomach with each swallow [1]. Gas within the stomach may exit via the esophagus by way of belching or via the duodenum, but most of the swallowed air should be regurgitated as a belch. The rate of excretion of gas per rectum is highly variable, ranging from 476 to 1,491 ml/day [2]. Gases are also produced while passing through the gastrointestinal tract, possibly resulting in abdominal symptoms although gas is continuously removed by eructation, anal evacuation, absorption through the intestinal mucosa, and bacterial consumption.

Esophagus

Gas passing through the esophagus may be derived from two sources: air swallowing and belching that is the passage of gaseous material from the stomach to the mouth. Nitrogen accounts for approximately 78% of swallowed air. Because little nitrogen is absorbed through the intestine and only about 400 ml of nitrogen is passed in flatus per day, most swallowed air should be eructated subconsciously or consciously [1]. It seems that 2–8 liters of air may be swallowed and eructated per day because of the 2–3 ml of air conveyed to the stomach with each swallow in an adult who usually consumes about 800 g of food and 1.5 liters of water.

Stomach and Duodenum

Since the esophagus enters the posterosuperior aspect of the stomach, gas is trapped above liquid overlying the gastroesophageal junction in the supine position. Therefore, most swallowed air is entrapped in the stomach and can travel along the small bowel.

In the upper gut, large quantities of gas are physiologically generated from chemical reactions of hydrogen ion and bicarbonate. For 1 mEq H^+ neutralized by bicarbonate in pancreatic, biliary, or duodenal secretions, 22.4 ml CO_2 is produced [3]. Fat digestion causes much gastric acid secretion and would result in an estimated CO_2 production rate of 1,400 ml/h. Most CO_2 produced during normal digestion is absorbed along the small bowel [4]. Thus, the luminal gas in the upper digestive tract is partly swallowed and in part originates from a series of chemical reactions of the foodstuffs within the gut. Theoretically, the calculated amount of CO_2 produced in the upper gut is 33.6 liters per day (1,400 ml multiplied by 24 h). However, the actual volume of CO_2 liberated as gas may be far below this value owing to the slow decomposition of

H_2CO_3 to CO_2 and H_2O in the absence of carbonic anhydrase and the relatively high solubility of CO_2 in water [5]. Presumably, approximately 5–10 liters of CO_2 seems to be produced and released into the lumen by chemical reactions.

As mentioned above, the pCO_2 rises dramatically in the duodenum, resulting in diffusion of CO_2 from lumen to blood, whereas CO_2 in swallowed air diffuses from the blood into the stomach [6]. Because the pO_2 of swallowed air is greater than that of blood in the stomach, O_2 is absorbed from the stomach. The pN_2 of swallowed air is slightly higher than that of venous blood in the stomach, possibly resulting in very slow absorption. In the duodenum, N_2 diffuses from blood into the lumen according to the partial pressure gradient because of the down gradient established by CO_2 production. In contrast, since pH_2 and pCH_4 are always higher in the lumen than in the blood, gases are constantly diffusing from lumen to blood. In our previous study using endoscopy, H_2 and CH_4 gases are more frequently detected in the stomach than expected, regardless of the presence of abdominal symptoms [7] although it has been reported that the stomach and duodenum harbor very low numbers of microorganisms adhering to the mucosal surface, typically less than 10^3 bacteria cells per gram of counts [8]. Fried et al. [9] reported that most of the bacteria identified from the duodenal aspirates belonged to species colonizing the oral cavity and pharynx, suggesting a descending route of colonization. Also, Thompson et al. [10] indicated that fermentation of ingested carbohydrate by oropharyngeal bacteria could contribute to measured breath hydrogen values soon after meal ingestion. Moreover, intraduodenal H_2 levels were higher in patients with severe atrophic gastritis than those without atrophic gastritis, and there was a progressive increase with the progression of atrophic gastritis [11]. In contrast, the intragastric H_2 level was the highest in patients with gastric mucosa of the closed type and was significantly higher than in those with severe atrophic gastritis. These results suggest that extensive atrophic gastritis may be more closely related to bacterial overgrowth in the jejunum, compared to that in the stomach.

Small and Large Intestine

There is a progressive increase in numbers of bacteria along the jejunum and ileum, from approximately 10^4 in the jejunum to 10^7 colony-forming units per gram of contents at the ileal end, with a predominance of Gram-negative aerobes and some obligate anaerobes [8]. In contrast, the large intestine is heavily populated by anaerobes and bacteria counts reach densities around 10^{12} colony-forming units per gram of luminal contents because transit time is slow and microorganisms have the opportunity to proliferate by fermenting available substrates derived from either the diet or endogenous secretions. Because bacteria represent the sole source of gut H_2 and CH_4, fasting breath H_2 and CH_4 gases have been used as markers of colonic fermentation [12, 13]. As H_2 production increases when a small amount of carbohydrate is supplied to colonic bacteria, the measurement of breath H_2 concentration has been proposed as an indicator of carbohydrate malabsorption [14]. H_2 gas is produced at a rate of 4 liters for every 12.5 g of undigested carbohydrate. Since it has been reported that 2–20% of carbohydrates escape small intestinal absorption [16] and men in their 40s consume 428 ± 72 g of carbohydrates [17], the calculated amount of H_2 produced in the colon is 2.7–27 liters/day (428 g × 0.2 × 4 liters/12.5 g = 27.4). Approximately 20% of all H_2 ingested or produced is eliminated via the lungs; the rest is either consumed or expelled via the rectum.

Several mechanisms of H_2 utilization have been reported in the human large intestine including methanogenesis, dissimilating sulfate reduction, and acetogenesis. The latter process corresponds to the reduction of 2 mol of CO_2 by 4 mol of H_2 to form 1 mol of acetate [18]. This process greatly decreases colonic gas volume. *Methanobrevibacter smithii*, which uses H_2 to reduce CO_2 to CH_4, is responsible for almost all the CH_4 produced in the intestine [19]. Since CH_4 production occurs primarily in the left colon whereas H_2 is produced primarily in the right colon [20], H_2 produced in the left colon may be rapidly converted to CH_4. CH_4 appears in the breath only when the numbers of methanogenic bacteria reach a critical level, about 10^8/g dry weight counts [19].

Clinical Problems

Irritable Bowel Syndrome

Intestinal gas is often incriminated as a major cause of irritable bowel syndrome (IBS), particularly when bloating is predominant, but no general relationship between intestinal gas and IBS symptoms has yet been established. Koide et al. [21] found a significantly increased intestinal gas volume score in plain abdominal radiographs in patients with IBS as compared with healthy controls, suggesting that abnormal accumulation of intestinal gas could be a problem in these patients. In contrast, Morken et al. [22] reported that intestinal gas volume is not correlated with abdominal discomfort after lactulose challenge and concluded that intestinal gas may not be the major cause of abdominal discomfort following carbohydrate ingestion in IBS.

In some diseases that favor bacterial proliferation, such as inflammatory bowel diseases, it may be difficult to determine the extent to which clinical deterioration is caused by bacterial overgrowth or the primary intestinal diseases. Actually, many individuals harbor bacterial overgrowth without symptoms. Thus, the clinical features of bacterial overgrowth, closely related to intraluminal gas production, vary greatly in individuals. However, several trials reported the results of H_2 breath testing for carbohydrate malabsorption and small intestinal bacterial overgrowth in patients with suspected IBS [23]. Intestinal fermentation products such as gas, short-chain fatty acids and intraluminal acidification could all affect colonic motility, possibly in part resulting in abdominal symptoms.

Functional Dyspepsia

Colonic fermentation influences not only the colonic function, but also the function of the upper digestive tract. Reduced proximal gastric tone after cecal infusion of lactulose and short-chain fatty acids [24] and relaxations of the lower esophageal sphincter [25] suggest that colonic fermentation might impair gastric accommodation to meals and induce gastroesophageal reflux. Several studies have shown that the presence of a large volume of air in the proximal stomach triggers stretch receptors in the gastric cardia and leads to an increase in the rate of transient relaxations of the lower esophageal sphincter. The distribution of gas in the digestive tract is easily detected by plain abdominal radiograph. In clinical practice, plain films of the abdomen are required in several different circumstances. The film of the abdomen with the patient standing erect is obtained to demonstrate fluid levels, a more detailed double-contrast view of the intestinal wall, or free air beneath the diaphragm. The other anteroposterior films with the patient lying on his back may corroborate findings on the erect study mainly to evaluate the distribution of the gas pattern of the abdomen. In the erect position, gas is retained largely in the fundus of the stomach. It has been reported that the volume and distribution of gas within the gut is closely associated with abdominal symptoms [26], whereas it has been unclear whether the gastric bubble on plain films is also linked to any symptom. The prevalence of GERD and dyspeptic symptoms were significantly lower in patients with a dome-type gastric bubble in our study [27]. Even a dome-type gastric bubble in the erect position may be distributed throughout the stomach in the supine position if a volume of gastric bubble is sufficient to fill the entire stomach. Changes in the form of gastric bubble detected on a plain film in the erect position may indicate the functional disorder of upper digestive tracts, such as decreased LES pressure, delayed gastric emptying, or impaired accommodation of the proximal stomach.

Intestinal Obstruction

The patient with colonic obstruction or delayed small intestinal transit may frequently have bacterial overgrowth and increased breath H2 levels because the bacterium can contact with food residues for a longer time. An informative case with ileus after local peritonitis is demonstrated in figure 2 [28]. A 70-year-old woman presented with abdominal pain and distention. She has a past history of surgical treatment for peritonitis before 13 years. Plain abdominal radiograph showed a

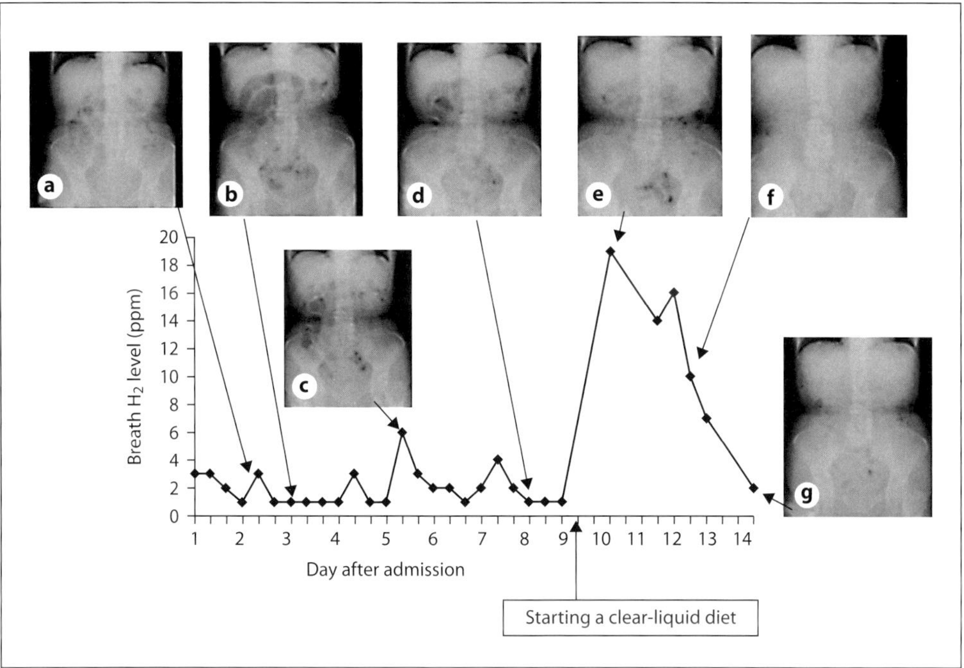

Fig. 1. Clinical course of a patient with ileus. Changes in breath hydrogen concentration and abdominal plain radiographs on representative dates are demonstrated.

markedly dilated small bowel, with no dilatation of the colon. Fasting breath H_2 concentration was 3 ppm at that time. This low levels of breath H_2 before treatment might indicate that unabsorbed food residues did not reach the cecum due to small bowel obstruction. The patient was treated with the use of a gastric tube for drainage of gastric juice. Nutrition was provided intravenously. During the first 4 days, although small bowel gas increased gradually (fig. 1a, b), breath H_2 concentrations remain less than 4 ppm (fig. 1). On the fifth day after admission, small bowel gas decreased and a small amount of colonic gas was demonstrated on the plain abdominal radiograph (fig. 1c). The breath H_2 concentration at that time increased from 1 to 6 ppm and reduced again to the baseline the next day. This change in breath H_2 level may possibly reflect the movement of intraluminal contents from the small intestine to the right colon. This value remained at a low level (1–2 ppm) from hospital days 5 through 8. Since intestinal gas was gradually reduced on radiographic imaging (fig. 1d), a liquid meal was supplied at noon on the ninth hospital day. The breath H_2 concentration on the next day increased markedly to 19 ppm despite that she had no abdominal symptoms and intestinal gas was not increased on the plain abdominal radiograph (fig. 1e). This suggests malabsorption of a liquid meal by which unabsorbed carbohydrates reach the colon and is utilized by fermentation, resulting in increased breath H_2 levels. Although

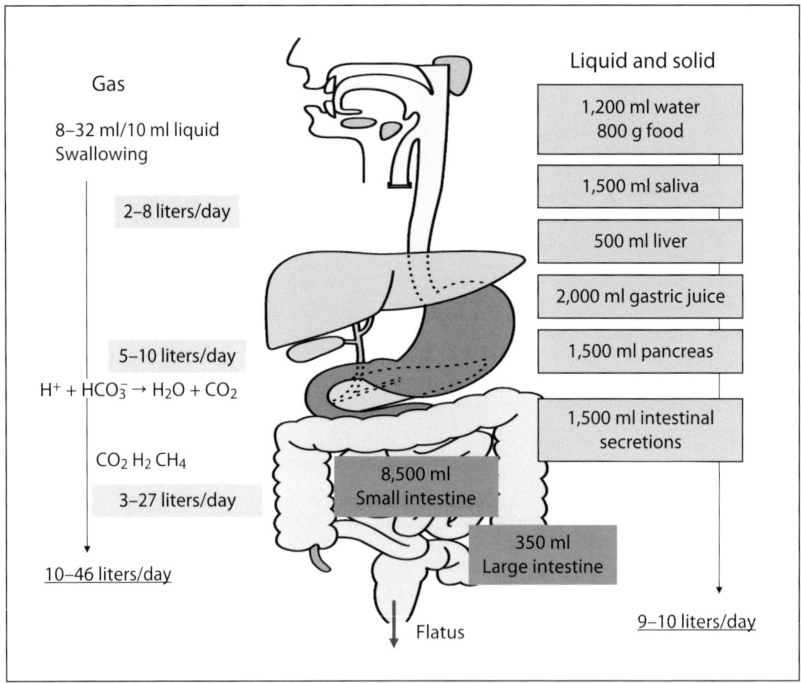

Fig. 2. Comparison of volume passing through the digestive tract between gas and liquid plus solid contents.

diets progressed from liquids to solids as tolerated, breath H_2 concentrations were decreased to 2 ppm 5 days after supplying meals. The patient did not complain of any abdominal symptoms during a routine progressive diet and intestinal gas concentration decreased continuously (fig. 1f, g). Digestive and absorptive function is considered to be restored gradually. Consecutive breath H_2 analysis could provide important information on the movement of intestinal food residues or the presence of malabsorption after meals.

Diverticulosis

High methanogen concentrations were found in patients with diverticulosis [29]. This suggests that the diverticula may provide an optimal environment for the growth of methanogens. It is possible that the diverticula may provide a sheltered niche where the slow-growing methanogens are not swept away and where symbiotic relationships with H_2-producing organisms may occur. Since CH_4 production occurs mainly in the left colon, H_2 produced in the left colon may be rapidly converted to CH_4, possibly resulting in reduced flatus.

Table 1. Direction of diffusion of gas between the lumen and mucosal blood

	N_2	O_2	CO_2	H_2	CH_4
Stomach		↓	↑		
Small intestine	↑		↓	↓	↓
Large intestine	↑	↑	↓	↓	↓

Upward arrow (↑) means the diffusion from blood to the lumen.
Downward arrow (↓) means the rapid absorption from the mucosa.

Conclusions

The sources of intraluminal gas are air swallowing, intraluminal production, and diffusion from the blood. Five gases (N_2, O_2, CO_2, H_2, and CH_4) account for more than 99% of gas passed per rectum. As shown in figure 2, greater amount of gases pass through the digestive tract per day than liquid and solid contents. There are geographical differences in gas metabolism (table 1). This suggests that impaired gas movement might be more closely associated with abdominal symptoms compared to liquid movement. Although it is unknown if increased and unequally distributed gases in the digestive tract are the key event in the pathogenesis of abdominal symptoms, intestinal gas metabolism might be one of the risk factors for developing various digestive diseases.

References

1 Maddock WG, Bell JL, Tremaine MJ: Gastrointestinal gas: observations on belching during anesthesia, operations and pyelography and rapid passage of gas. Ann Surg 1949;130:512–537.

2 Tomlin J, Lowis C, Read NW: Investigation of normal flatus production in healthy volunteers. Gut 1991;32:665–669.

3 Suarez F, Levitt MD: Intestinal gas; in Feldman M, Friedman LS, Sleisenger MH (eds): Sleisenger and Fordtran's Gastrointestinal and Liver Disease: Pathophysiology/Diagnosis/Management. Philadelphia, Saunders, 2002, pp 155–163.

4 Fordtran JS, Walsh JH: Gastric acid secretion rate and buffer content of the stomach after eating: results in normal subjects and in patients with duodenal ulcer. J Clin Invest 1973;52:645–647.

5 Fordtran JS, Morawski SG, Santa Ana CA, Rector FC Jr: Gas production after reaction of sodium bicarbonate and hydrochloric acid. Gastroenterology 1984;87:1014–1121.

6 Levitt MD: Volume and composition of human intestinal gas determined by means of an intestinal washout technique. N Engl J Med 1971;284:1394–1398.

7 Urita Y, Ishihara S, Akimoto T, Kato H, Hara N, Honda Y, Nagai Y, Nakanishi K, Shimada N, Sugimoto M, Miki K: Hydrogen and methane gases are frequently detected in the stomach. Word J Gastroenterol 2006;12:3088–3091.

8 Neish AS: Microbes in gastrointestinal health and disease. Gastroenterology 2009;136:65–80.

9 Fried M, Siegrist H, Frei R, Froehlich F, Duroux P, Thorens J, Blum A, Bille J, Gonvers JJ, Gyr K: Duodenal bacterial overgrowth during treatment in outpatients with omeprazole. Gut 1994;35:23–26.

10 Thompson DG, O'Brien JD, Hardie JM: Influence of the oropharyngeal microflora on the measurement of exhaled breath hydrogen. Gastroenterology 1986; 91:853–860.

11 Urita Y, Watanabe T, Maeda T, Sasaki Y, Ishii T, Yamamoto T, Kugahara A, Nakayama A, Nanami M, Domon K, Ishihara S, Kato H, Hike K, Sanaka M, Nakajima H, Sugimoto M, Miki K: Extensive atrophic gastritis linked to increased levels of intraluminal hydrogen gas. Hepatogastroenterology 2008;55:1645–1648.
12 Le Marchand L, Wilkens LR, Harwood P, Cooney RV: Use of breath hydrogen and methane as markers of colonic fermentation in epidemiologic studies; circadian patterns of excretion. Environ Health Perspect 1992;98:199–202.
13 Perman JA, Modler S, Barr RG, Rosenthal P: Fasting breath hydrogen concentration: normal values and clinical application. Gastroenterology 1984;87: 1358–1363.
14 Levitt MD, Donaldson RM: Use of respiratory hydrogen (H_2) excretion to detect carbohydrate malabsorption. J Lab Clin Med 1970;75:937–945.
15 Strocchi A, Levitt MD: Intestinal gas; in Feldman M, Scharschmidt BF, Sleisenger MH (eds): Sleisenger and Fordtran's Gastrointestinal and Liver Disease, ed 6. Philadelphia, Saunders, 1998, vol 1, p 155.
16 Stephen AM, Haddad AC, Philipps SF: Passage of carbohydrate into the colon: direct measurements in humans. Gastroenterology 1983;85:589–595.
17 Nakamura T, Takebe K, Tando Y, Arai Y, Yamada N, Ishii M, Kikuchi H, Machida K, Imamura K, Terada A: Serum fatty acid composition in normal Japanese and its relationship with dietary fish and vegetable oil contents and blood lipid levels. Ann Nutr Metab 1995;39:261–270.
18 Gibson GR, Cummings JH, Macfarlane GT, Allison C, Segal I, Vorster HH, Walker AR: Alternative pathways for hydrogen disposal during fermentation in the human colon. Gut 1990;31:679–683.
19 Miller TL, Wolin MJ: Enumeration of *Methanobrevibacter smithii* in human feces. Arch Microbiol 1982;131:14–18.
20 Bond JH Jr, Engel RR, Levitt MD: Factors influencing pulmonary methane excretion in man. An indirect method of studying the in situ metabolism of the methane-producing colonic bacteria. J Exp Med 1971;133:572–588.
21 Koide A, Yamaguchi T, Odaka T, Koyama H, Tsuyuguchi T, Kitahara H, Ohto M, Saisho H: Quantitative analysis of bowel gas using plain abdominal radiograph in patients with irritable bowel syndrome. Am J Gastroenterol 2000;95:1735–1741.
22 Morken MH, Berstad AE, Nysaeter G, Berstad A: Intestinal gas in plain abdominal radiographs does not correlate with symptoms after lactulose challenge. Eur J Gastroenterol Hepatol 2007;19:589–593.
23 Cash BD, Chey WD: Irritable bowel syndrome, an evidence-based approach to diagnosis. Aliment Pharmacol Ther 2004;19:1235–1245.
24 Ropert A, Cherbut C, Roze C, Le Quellec A, Holst JJ, Fu-Cheng X, des Varannes SB, Galmiche JP: Colonic fermentation and proximal gastric tone in humans. Gastroenterology 1996;111:289–296.
25 Piche T, des Varannes SB, Sacher-Huvelin S, Holst JJ, Cuber JC, Galmiche JP: Colonic fermentation influences lower esophageal sphincter function in gastroesophageal reflux disease. Gastroenterology 2003;124:894–902.
26 Harder H, Serra J, Azpiroz F, Passos MC, Aguade S, Malagelada JR: Intestinal gas distribution determines abdominal symptoms. Gut 2003;52:1708–1713.
27 Watanabe T, Urita Y, Maeda T, Sasaki Y, Hike K, Sanaka M, Nakajima H, Sugimoto M: Form of gastric air bubble is associated with gastroesophageal reflux symptoms. Hepatogastroenterology 2009;56: 1566–1570.
28 Urita Y, Watanabe T, Maeda T, Sasaki Y, Ishihara S, Hike K, Sanaka M, Nakajima H, Sugimoto M: Breath hydrogen gas concentration linked to intestinal gas distribution and malabsorption in patients with small-bowel pseudo-obstruction. Biomarker Insights 2009;4:9–15.
29 Weaver GA, Krause JA, Miller TL, Wolin MJ: Incidence of methanogenic bacteria in a sigmoidoscopy population: an association of methanogenic bacteria and diverticulosis. Gut 1986;27:698–704.

Yoshihisa Urita
Department of General Medicine and Emergency Care, Toho University, School of Medicine
6–11–1, Omori-Nishi, Ota-Ku
Tokyo, 143-8541 (Japan)
Tel. +81 3 3762 4151, Fax +81 3 3765 6518, E-Mail foo@eb.mbn.or.jp

Yoshikawa T, Naito Y (eds): Gas Biology Research in Clinical Practice.
Basel, Karger, 2011, pp 15–23

Therapeutic Medical Gas

Atsunori Nakao

Department of Surgery, University of Pittsburgh Medical Center, Pittsburgh, Pa., USA

Abstract

Therapeutic medical gas is pharmaceutical gaseous molecules which offer solutions to medical needs. In addition to traditional medical gases including oxygen and nitrous oxide, a number of medical gases have been recently discovered to play protective roles in various disease conditions. In particular, nitric oxide, carbon monoxide and hydrogen sulfide are found to be endogenously generated in the human body and mediate signaling pathways as biological messengers, and are shown to have potent cytoprotective effects. Herein, we summarize the recent advances of therapeutic medical gas research and discuss their clinical feasibility, typically such as nitric oxide, carbon monoxide, hydrogen sulfide, hydrogen, xenon, helium and ozone. A recent increase in publications in the medical gas field clearly indicates that there are significant opportunities for the use of medical gases as therapeutic tools. It would be necessary to identify safety concerns when using therapeutic gases along with potential side effects and toxicity. Although further investigations are required, medical gas may provide a huge impact as a novel and innovative therapeutic tool for unmet medical needs with considerable health burdens.

What Is 'Therapeutic Medical Gas'?

The definition of 'therapeutic medical gas' is the pharmaceutical gaseous molecules which are applied to the human body for the purpose of a variety of medical conditions and producing other desirable effects. It is no doubt that oxygen therapy to increase tissue oxygenation has been a mainstream in critical care medicine. However, recent observations in experimental and clinical studies clearly revealed the needs to provide additional supplemental gases to patients with many etiologies. The list of therapeutic medical gases has been growing continuously (table 1). Herein, we discuss the recent advances in medical gas research and delivery mechanisms including clinical application for special gases with recently discovered roles as protective properties.

Table 1. List of recent therapeutic medical gases

	Nitric oxide	Carbon monoxide	Hydrogen sulfide	Hydrogen	Xenon	Helium	Ozone
Formula	NO	CO	H_2S	H_2	Xe	He	O_3
Color, odor	colorless, a mild, sweet odor	colorless, odorless	colorless, smell like rotten eggs	colorless, odorless	colorless, odorless	colorless, odorless	pale blue a sharp, cold, irritating odor
Flammable	no	no	yes	yes	yes	no	no
Toxicity	yes	yes	yes	no	yes	no	yes
Produced in mammalian cells?	yes; from L-arginine by nitric oxide synthase (NOS), or reduction of nitrite	yes; through heme degradation by heme oxygenase	yes; from L-cysteine by CBS and CBE	no	no	no	yes; in the white blood cells and other biological systems
Effects for vessels	vasodilation	vasodilation	vasodilation	no change	no change	vasodilation	vasodilation
Anti-apoptotic effects	yes	yes	yes	yes	yes	not shown	yes
Anti-inflammatory effects	yes	yes	yes	yes	yes	yes	yes

Advantage of Medical Gas Therapy

Gas therapy has distinct advantages over pharmaceutical drugs; it easily penetrates biomembranes and diffuses into the cytosol, mitochondria and nucleus to reach target tissues. Therefore, therapeutic medical gases may be used for a variety of disorders in various clinical settings [1]. Medical gases can be administered in a straightforward way simply by providing the gas for the patients to inhale using a ventilator circuit, facemask, or nasal cannula. In most cases, administration of medical gas by inhalation may not complicate the current existing therapeutic strategies by simply adding the gas in the conventional delivery mechanism. The ability to administer medical gas inhalationally makes it extremely attractive from a feasibility standpoint for translation into a human clinical setting. Some gases such as supplemental oxygen may be administered in the patient's home or in a medical setting. However, safety warrants uses of other therapeutic gases are given under the direct care of a physician and supervision of trained medical staff.

Handling and Delivery of Therapeutic Medical Gases

As these medical gases may be toxic, hazardous or poisonous at a higher concentration, methods and devices for the delivery of therapeutic gases must be carefully examined. The standard yoking system with compressed gas cylinder valve outlet and inlet connections with specific regulators will help avoid improper gas therapy delivery. The best delivered system will be through a closed system in which the delivery mechanism is not prone to problematic leaks or air entrainment. All techniques using gas therapy have the obligation to identify by label the different available therapy gases before administration, and provide safe gas delivery and precision gas analysis or monitoring.

In addition to the development of safe devices for inhaled medical gas, potential clinical application may include a parenteral injectable or a drug containing a gas-releasing moiety. Emerging evidence reveals the efficacies of a novel class of substance, CO-releasing molecules (CORMs), which are capable of exerting a variety of pharmacological activities via the liberation of controlled amounts of CO in biological systems [2]. Recent studies show that nitrate and nitrite, contained in green vegetables, can be recycled in vivo to form NO, representing an important alternative source of NO to the classical L-arginine-NO-synthase pathway, in particular in hypoxic states [3]. It is noteworthy that the enterobacterial flora is an important source of hydrogen and hydrogen sulphide. Experimental study has demonstrated that when hydrogen is released by intestinal bacteria systemic antibiotics may affect the stable levels of these gases along with suppression of commensal bacteria, which may influence to host's resistance and innate immunity [4].

Application of Medical Gas for Disease

Medical research must be ultimately conducted for human health benefits. Possible therapeutic opportunities for medical gases are shown in table 1. Although every experimental success might not be transferable to standard clinical practice in the near future, the sophisticated experimental concepts of gas inhalation therapy for a medical condition can be considered to be an important step toward clinical application. The medical gas research is relatively unexplored with a short history. Appropriately designed randomized controlled trials with patient-important outcomes, such as improvement of functions of target organs, decreased intensive care unit and hospital days, and decreased cost of therapy, are sorely needed to establish the role of medical gas therapy in patients with disease, associated with the benefits over preexisting standard therapy. At this moment, INOmax®, NO gas for inhalation therapy, is a United States Food and Drug Administration (FDA)-approved drug for the treatment of hypoxic respiratory failure in term and near-term newborns. The drug is also approved by regulatory authorities and used in the clinical setting in Europe,

Australia, Asia and Latin America. A European clinical study revealed that inhalation of 100–125 ppm CO by patients with chronic obstructive pulmonary disease in a stable phase was feasible and led to trends in reduction of sputum eosinophils and improvement of responsiveness to methacholine [5]. In the USA, a single-blind, placebo-controlled, dose-escalating phase 2 study of inhaled CO in patients receiving renal transplants is currently conducted using Covox®, a device for CO inhalation. The primary endpoint of the study is to evaluate the safety and tolerability of increasing CO dose levels when administered as an inhaled gas to kidney transplant patients over the course of 1 h in an acute hospital setting. No matter how much time and money are spent, the successful development of a medical gas therapy, as a new therapeutic tool, is by no means a certainty. Researchers and clinicians should be dedicating effort into the clinically feasible use of medical gas that fulfills the unmet medical needs at the frontline of healthcare.

Nitric Oxide

Nitric oxide (NO) is a colorless and poisonous gas which is generated by automobile and thermal power plants and causes a serious air pollutant. NO concentration in unpolluted air is approximately 0.01 parts per million (ppm). However, NO is an important signaling molecule in the body of mammals and was named 'Molecule of the Year' in 1992 [6]. NO plays an important role in vascular homeostasis by its potent vasoregulatory and immunomodulatory properties. Blood vessel dilation is one of the most well-known effects of NO. NO stimulates soluble guanylate cyclase (sGC) and increases cGMP content in vascular smooth muscle cells, resulting in relaxation of vascular tone and vasodilation. In addition to its vasorelaxant effect, NO has a complex spectrum of actions including the regulation of platelet activity and the preservation of the normal structure of the vessel wall. Thus, the actions of NO on blood vessels may increase tissue blood supply and abate the inflammatory response, leading to protection of the tissues from oxidative insults. Inhaled NO is already in clinical use for the treatment of hypoxic respiratory failure and pulmonary hypertension particularly in neonates and additional uses of NO are being investigated in other settings of lung and cardiac diseases [7]. Although multiple single-center studies demonstrated the ability of inhaled NO to improve the outcome of patients with adult ARDS were marginal [8], some studies advocate inhalation of NO as a method to prevent graft injury due to ischemia/reperfusion injury after human lung and liver transplantation.

Carbon Monoxide

Carbon monoxide (CO) is an invisible, chemically inert, colorless and odorless gas and is commonly viewed as an environmental pollutant associated with toxic effects resulting from its ability to compete with oxygen for binding to hemoglobin. CO avidly binds to hemoglobin and forms carboxyhemoglobin (COHb) with an affinity 240 times higher than that of oxygen, resulting in interference with the oxygen-carrying capacity of the blood and consequent tissue hypoxia. Recent basic research has revealed that endogenous CO is an important physiological regulatory factor and exerts anti-inflammatory, anti-apoptotic and organ/cellular protective effects. CO is endogenously and physiologically generated in mammalian cells via the catabolism of heme in the rate-limiting step by heme oxygenase (HO) systems [9]. Potent therapeutic efficacies of CO have been demonstrated using experimental models for many conditions, including paralytic ileus, hemorrhagic shock [10], hyperoxic lung injury, and endotoxemia, supporting the new paradigm that, at low concentrations, CO functions as a signaling molecule that exerts significant cytoprotection. Similarly, bacteria are associated with at least one-third of COPD exacerbations. Toll-like receptors (TLRs) are needed for recognition and clearance of bacteria. In macrophages, CO has recently been shown to inhibit signaling by TLR2, TLR4, TLR5 and TLR9 (but not TLR3) [11]. Soluble forms of CO, such as CO-releasing molecules, may overcome the problem of toxicity and allow clinical application [2].

Hydrogen Sulfide

Hydrogen sulfide (H_2S) is a colorless, toxic and flammable gas. It is a naturally occurring gas found in volcanic gases and some well waters and is also responsible for the foul odor of rotten eggs and flatulence. The toxic effects of H_2S in humans include eye irritation, shortness of breath, and chest tightness at concentrations <100 ppm. Exposure to H_2S at >1,000 ppm may cause severe adverse effects, ranging from loss of consciousness to fatality. H_2S is endogenously synthesized normally in vertebrates from L-cysteine, a product of food-derived methionine, by the cystathionone-β-synthase (CBS) and cyctathione-γ-lyase (CSE) system. The enterobacterial flora is another source of H_2S. H_2S is believed to help regulate body temperature and metabolic activity at physiological concentrations [12]. Also, H_2S exerts physiological effects in the cardiovascular system, possibly through modulation of K^+-ATP channel opening or as a cellular messenger molecule involved in vascular flow regulation [13]. Administration of H_2S produced a 'suspended animation-like' metabolic status with hypothermia and reduced oxygen demand [14], thus protecting from lethal hypoxia. This hypometabolic state, which resembles hibernation, induced by H_2S may contribute to tolerance against oxidative stress. The effects of a soluble form of H_2S (using sodium sulfide) have been under investigation for clinical study on the patients who underwent coronary artery bypass graft (CABG) to potentially reduce the damage done to the heart during surgery.

Hydrogen

Hydrogen (H_2) is the lightest and most abundant of chemical elements, constituting nearly 90% of the universe's elemental mass. In contrast, earth's atmosphere contains less that 1 ppm of hydrogen. H_2 is known to be highly flammable and to violently react with oxidizing elements as typified by the 1937 Hindenberg Zeppelin disaster. Therefore, the use of H_2 as a therapeutic agent is not intuitively obvious. However, recent evidence indicates that inhaled H_2 gas has antioxidant and anti-apoptotic properties that can protect organs from ischemia-reperfusion-induced injury by selectively scavenging detrimental ROS. The mechanism of action of inhaled H_2 gas in these models involves its ability to prevent oxidative damage, as indicated by decreased nucleic acid oxidation and lipid peroxidation [15, 16]. H_2-rich liquid such as H_2 water represents a novel and easily translatable method of delivery of molecular H_2. H_2 water may be of potential therapeutic value in the treatment of oxidative stress-induced pathologies as well as inhaling H_2 gas [17, 18]. A clinical trial in type 2 diabetic patients given supplemental H_2 water led to improved lipid and glucose metabolism compared to controls [19].

Xenon

Xenon is a colorless, odorless noble gas considered chemically inert and unable to form compounds with other molecules. Xenon is a trace gas in Earth's atmosphere, occurring at $<0.087 \pm 0.001$ ppm and is also found in gases emitted from some mineral springs. Since Cullen and Gross [20] first used xenon on human patients in 1951, xenon has been successfully used in a number of surgical operations as an anesthetic agent. Xenon readily crosses the blood-brain barrier and has low blood/gas solubility, which is advantageous for rapid inflow and washout, associated with good cardiovascular stability and satisfactory sedation [21]. In addition to its anesthetic properties, xenon has protective effects against cerebral ischemia [22]. Decreased blood flow to the brain leads to neuronal death through necrotic and apoptotic mechanisms, which are largely dependent on the activation of the N-methyl-D-aspartate (NMDA) receptor. Since xenon effectively inhibits the NMDA receptor, the neuroprotective effects of xenon may be at least partially due to this inhibition [22–25]. There is evidence suggesting that brief exposure to xenon prevents myocardial ischemia/reperfusion injury.

Helium

Helium is an odorless, tasteless, nonexplosive, noncombustible, nontoxic and physiologically inert gas and has been used safely in the treatment of respiratory disease since 1979 [26, 27]. It diffuses very rapidly to the target organs because of its low atomic weight. It is also nonreactive with body tissues and relatively insoluble in body fluid. The physical properties of helium such as its low density, laminar flow and low driving pressure decrease airway resistance making it an attractive treatment for respiratory diseases that involve a decrease in airway diameter and consequently increased airway resistance, such as bronchial asthma, chronic obstructive pulmonary disease and bronchiolitis. Patients with these conditions may suffer a range of symptoms including breathlessness, hypoxemia and eventually a weakening of the respiratory muscles, which can lead to respiratory failure requiring intubation and mechanical ventilation. Helium can reduce all these effects, making it easier for the patient to breathe by lowered airway resistance, thereby requiring less mechanical energy to ventilate the lungs. Helium is reported to increase in the coronary collateral circulation and enhance the vasodilator effect of inhaled nitric oxide on pulmonary vessels. The physical properties of helium may result in more efficient uptake oxygen and a facilitation of nitrogen egress from affected mitochondria in transient cerebral ischemia, both of which may explain its beneficial effects.

Ozone

Ozone is a triatomic molecule consisting of three oxygen atoms. Ozone is a pale blue gas with a sharp, cold, irritating odor and is produced naturally by electrical discharges following thunderstorms or ultraviolet (UV) rays emitted from the sun. Ozone is present in low concentrations throughout the Earth's atmosphere; however, an ozone layer exists between 10 and 50 km above from the surface of the earth and plays a very important role filtering UV rays which is critical for the maintenance of biological balance in the biosphere. Ozone gas has a high oxidation potential and is widely used in treatment of water in aquariums and fish ponds to minimize bacterial growth, control parasites, and eliminate transmission of some diseases. Ozone inhalation (0.1–1 ppm) can be toxic to the pulmonary system and cause upper respiratory irritation, rhinitis, headache, and occasionally nausea and vomiting. However, ozone has been used as a therapeutic agent for the treatment of different diseases by creating resistance against oxidative stress via inducing an antioxidative system. Ozone, administered by rectal insufflation, prior to ischemia/reperfusion injury, prevents the damage induced by ROS and attenuates renal and hepatic ischemia/reperfusion injury [28]. Medical applications of blood ozonation via extracorporeal blood oxygenation and ozonation was found to be safe and effective in treating peripheral artery disease in clinical trials [29, 30].

Conclusions

As highlighted above, the ability of medical gases to ameliorate oxidative stress plays an important role at the chemical, cellular and physiological levels. Although some medical gases may cause serious adverse effects, there are still many possible applications of these gases as therapeutic tools for various diseases if the concentrations are tightly controlled. The future of medical gas therapy must focus on the establishment of safe and well-defined administration parameters and on randomized controlled trials to determine the precise indications and guidelines for the use of medical gases in the treatment of various pathologies.

References

1 Nakao A, Sugimoto R, Billiar TR, et al: Therapeutic antioxidant medical gas. J Clin Biochem Nutr 2009;44:1–13.

2 Motterlini R, Mann BE, Foresti R: Therapeutic applications of carbon monoxide-releasing molecules. Expert Opin Investig Drugs 2005;14:1305–1318.

3 Lundberg JO, Weitzberg E, Gladwin MT: The nitrate-nitrite-nitric oxide pathway in physiology and therapeutics. Nat Rev Drug Discov 2008;7:156–167.

4 Kajiya M, Sato K, Silva MJ, et al: Hydrogen from intestinal bacteria is protective for concanavalin A-induced hepatitis. Biochem Biophys Res Commun 2009;386:316–321.

5 Bathoorn E, Slebos DJ, Postma DS, et al: Anti-inflammatory effects of inhaled carbon monoxide in patients with COPD: a pilot study. Eur Respir J 2007;30:1131–1137.

6 Culotta E, Koshland DE Jr: NO news is good news. Science1992;258:1862–1865.

7 Creagh-Brown BC, Griffiths MJ, Evans TW: Bench-to-bedside review: Inhaled nitric oxide therapy in adults. Crit Care 2009;13:221.

8 Michael JR, Barton RG, Saffle JR, et al : Inhaled nitric oxide versus conventional therapy: effect on oxygenation in ARDS. Am J Respir Crit Care Med 1998;157:1372–1380.

9 Tenhunen R, Marver HS, Schmid R: The enzymatic conversion of heme to bilirubin by microsomal heme oxygenase. Proc Natl Acad Sci USA 1968;61:748–755.

10 Pannen BH, Kohler N, Hole B, et al: Protective role of endogenous carbon monoxide in hepatic microcirculatory dysfunction after hemorrhagic shock in rats. J Clin Invest 1998;102:1220–1228.

11 Nakahira K, Kim HP, Geng XH, et al: Carbon monoxide differentially inhibits TLR signaling pathways by regulating ROS-induced trafficking of TLRs to lipid rafts. J Exp Med 2006;203:2377–2389.

12 Kamoun P: Endogenous production of hydrogen sulfide in mammals. Amino Acids 2004;26:243–254.

13 Leslie M: Medicine. Nothing rotten about hydrogen sulfide's medical promise. Science 2008;320:1155–1157.

14 Blackstone E, Morrison M, Roth MB: H_2S induces a suspended animation-like state in mice. Science 2005;308:518.

15 Ohsawa I, Ishikawa M, Takahashi K, et al: Hydrogen acts as a therapeutic antioxidant by selectively reducing cytotoxic oxygen radicals. Nat Med 2007;13:688–694.

16 Hayashida K, Sano M, Ohsawa I, et al: Inhalation of hydrogen gas reduces infarct size in the rat model of myocardial ischemia-reperfusion injury. Biochem Biophys Res Commun 2008;373:30–35.

17 Ohsawa I, Nishimaki K, Yamagata K, et al: Consumption of hydrogen water prevents atherosclerosis in apolipoprotein E knockout mice. Biochem Biophys Res Commun 2008;377:1195–1198.

18 Cardinal JS, Zhan J, Wang Y, et al: Oral administration of hydrogen water prevents chronic allograft nephropathy in rat renal transplantation. Kidney Int 2010;77:101–109.

19 Kajiyama S, Hasegawa G, Asano M, et al: Supplementation of hydrogen-rich water improves lipid and glucose metabolism in patients with type 2 diabetes or impaired glucose tolerance. Nutr Res 2008;28:137–143.

20 Cullen SC, Gross EG: The anesthetic properties of xenon in animals and human beings, with additional observations on krypton. Science 1951;113:580–582.

21 Bedi A, Murray JM, Dingley J, et al: Use of xenon as a sedative for patients receiving critical care. Crit Care Med 2003;31:2470–2477.

22 David HN, Haelewyn B, Rouillon C, et al: Neuroprotective effects of xenon: a therapeutic window of opportunity in rats subjected to transient cerebral ischemia. Faseb J 2008;22:1275–1286.

23 Franks JJ, Horn JL, Janicki PK, et al: Halothane, isoflurane, xenon, and nitrous oxide inhibit calcium ATPase pump activity in rat brain synaptic plasma membranes. Anesthesiology 1995;82:108–117.

24 Ma D, Wilhelm S, Maze M, et al: Neuroprotective and neurotoxic properties of the 'inert' gas, xenon. Br J Anaesth 2002;89:739–746.
25 Wilhelm S, Ma D, Maze M, et al : Effects of xenon on in vitro and in vivo models of neuronal injury. Anesthesiology 2002;96:1485–1491.
26 Hess DR, Fink JB, Venkataraman ST, et al: The history and physics of heliox. Respir Care 2006;51:608–612.
27 Pan Y, Zhang H, VanDeripe DR, et al: Heliox and oxygen reduce infarct volume in a rat model of focal ischemia. Exp Neurol 2007;205:587–590.
28 Ajamieh H, Merino N, Candelario-Jalil E, et al: Similar protective effect of ischaemic and ozone oxidative preconditioning in liver ischaemia/reperfusion injury. Pharmacol Res 2002;45:333–339.
29 Di Paolo N, Bocci V, Gaggiotti E: Ozone therapy. Int J Artif Organs 2004;27:168–175.
30 Bocci V: Ozone as Janus: this controversial gas can be either toxic or medically useful. Mediators Inflamm 2004;13:3–11.

Dr. Atsunori Nakao
Department of Surgery, University of Pittsburgh Medical Center
Pittsburgh, PA 15213 (USA)
Tel. +1 412 648 9547, Fax +1 412 624 6666, E-Mail anakao@imap.pitt.edu

Yoshikawa T, Naito Y (eds): Gas Biology Research in Clinical Practice.
Basel, Karger, 2011, pp 24–34

Analysis of Breath CO and Application to Hemodynamic Monitoring

Makoto Sawano

Saitama Medical University, Kamoda, Kawagoe, Saitama, Japan

Abstract

Breath tests have replaced more invasive conventional tests in the diagnosis of digestion and absorption disorders or gastrointestinal infection. However, highly invasive techniques are still widely used for hemodynamic monitoring. Although less invasive replacement of these techniques has long been desired, no breath test has succeeded in this field, because of its difficulty in continuous measurement of rapidly fluctuating parameters. Thermodilution using a pulmonary-artery catheter is the gold standard in the estimation of cardiac output. However, recent trials showed an increased mortality in the group of cardiac failure patients treated using a pulmonary-artery catheter. There is no question that cardiac output estimation provides useful information for management of circulatory disorders, undesired outcome should be attributed to high incidence and severity of complications associated with pulmonary-artery catheter placement. Radioisotope-labeled red-cell dilution is the gold standard in estimation of circulating blood volume. Carboxyhemoglobin dilution was developed as a possible replacement of the radioisotope dilution. However, infusion of 100 ml carbon monoxide saturated autologous blood was required and was potentially hazardous to patients under critical conditions. The author developed accurate, noninvasive, continuous carboxyhemoglobin densitometry by expired gas analysis. Application of this technique to low-dose carboxyhemoglobin dilution achieved minimally invasive estimation of cardiac output and circulating blood volume. The only prerequisite for the estimation is the infusion of 20 ml CO-saturated autologous blood via the peripheral vein. Hemodynamic monitoring by expired gas analysis achieved good agreement with conventional methods and is acceptable as an equivalently accurate but low invasive replacement.

In the past decade, breath tests, in combination with stable isotope tracers, have achieved remarkable success in replacing more invasive conventional tests in the fields such as assessment of digestion and absorption functions or diagnosis of gastrointestinal infection. However, highly invasive conventional techniques are still widely used for measuring cardiac output and circulating blood volume, the two basic parameters of hemodynamic monitoring. Although less invasive replacement of these techniques has long been desired, no breath test succeeded in this field so far, because of its difficulty in continuous measurement of rapidly fluctuating parameters [1].

Thermodilution using a pulmonary-artery catheter (PAC) is the gold standard in the estimation of cardiac output. However, recent randomized controlled trials showed increased mortality in the group of cardiac failure patients treated using a PAC [2–5]. There is no question that cardiac output estimation provides useful and sometimes decisive information for the management of circulatory disorders, and an undesired outcome should be attributed to the high incidence and severity of complications associated with PAC placement [6].

Radioisotope-labeled red blood cell (^{51}Cr-RBC or ^{99m}Tc-RBC) dilution is generally accepted as the gold standard of red cell volume or circulating blood volume measurement [7, 8]. However, this method has not been commonly used in clinical applications, due to restrictions in use of radioactive isotopes at the bedside. Carboxyhemoglobin dilution was introduced as a replacement for ^{51}Cr-RBC. The accuracy of carboxyhemoglobin dilution has been shown to be equivalent to that of the ^{51}Cr-RBC method [9–11], but it requires a bolus infusion of 100 ml of carbon monoxide (CO)-saturated blood, which is potentially hazardous for anemic or hypovolemic patients [12]. Ensuring the safety of this method requires reduction of the infusion volume, which necessitates development of techniques with higher resolution in carboxyhemoglobin fraction (COHb%) measurement.

The author developed a novel technique of breath test, namely the reference gas method, which enabled accurate, noninvasive and continuous carboxyhemoglobin densitometry in the pulmonary circulation [12]. Application of this technique to low-dose carboxyhemoglobin dilution achieved a minimally invasive estimation of cardiac output and circulating blood volume. The only prerequisite for the estimation is the infusion of 20 ml CO-saturated autologous blood via the peripheral vein.

In this article, the author first introduces the procedures and algorithms of carboxyhemoglobin densitometry using expired gas analysis (the EGA method) and its application to cardiac output and circulating blood volume estimation by low-dose carboxyhemoglobin dilution. Then, the agreements between the estimations by EGA and those by conventional methods are presented to assess whether EGA is acceptable as an equivalently accurate but low invasive replacement. Agreements between the two methods were assessed statistically using the Bland-Altmann method [13].

Procedures for Hemodynamic Monitoring by EGA

Preparation of the CO-Saturated Autologous Blood

Draw 25 ml of venous blood, and determine the baseline carboxyhemoglobin concentration (COHb%) using a CO-hemoximeter with a resolution of 0.1%. Bubble the blood with 100% CO gas in a closed chamber continuously for approximately 20 min. This process usually achieves COHb% over 90%, which is confirmed using the CO-hemoximeter.

Continuous Determination of Expired Gas Carbon Monoxide and Carbon Dioxide Concentrations

Inspired and expired gases are continuously sampled from the ventilator circuit or the mask with 2 one-way valves to determine CO and CO_2 concentrations using a Carbolizer (mba2000, Taiyo, Osaka, Japan). The Carbolizer is capable of continuous determination of CO and CO_2 concentrations every second, with resolutions of 0.1 ppm and 0.1%, respectively. The zero-point calibration for CO and CO_2 is performed using the inspired gas prior to the measurements.

Expired gas CO and CO_2 concentrations are measured for a minute to determine the baseline. Then, following a bolus infusion of 20 ml CO-saturated autologous blood via a peripheral vein, expired gas CO and CO_2 concentrations are continuously measured for 6 min. 5 min 30 s after the infusion, a small amount (1 ml) of venous blood is drawn and COHb% is determined using the CO-hemoximeter.

Estimation of End-Tidal CO from Expiratory Gas CO and CO_2 (Reference Gas Method)

Equation (1) expresses the relation between end-tidal CO concentration (ETCO), end-tidal CO_2 concentration ($ETCO_2$), expiratory gas CO and CO_2 concentrations at a random respiratory phase (exCO, $exCO_2$) and inspiratory gas CO and CO_2 concentrations (inCO, in CO_2):

$$\frac{(\text{exCO}-\text{inCO})}{(\text{exCO}_2-\text{inCO}_2)} = \frac{(\text{ETCO}-\text{inCO})}{(\text{ETCO}_2-\text{inCO}_2)} \quad (1)$$

The zero-point calibration of the Carbolizer using inspiratory gas allows simplification of equation (1) by assigning the inCO and $inCO_2$ value of 0 kPa, as follows:

$$\frac{\text{exCO}}{\text{exCO}_2} = \frac{\text{ETCO}}{\text{ETCO}_2} \quad (2)$$

$ETCO_2$ (or alveolar CO_2 concentration) is calculated from arterial CO_2 tension ($PaCO_2$), atmospheric tension (P_{atm}), and saturated water vapor tension (P_{vap}). As CO_2 is highly diffusible, difference between alveolar and capillary CO_2 tensions can be ignored. Equation (2) is solved for ETCO by assigning a P_{atm} value of 101 kPa and a P_{vap} value of 6.27 kPa (standard values for temperature of 37.0°C) as follows:

$$ETCO=0.011 \times PaCO_2 \times \frac{exCO}{exCO_2} \tag{3}$$

Estimation of COHb% from ETCO

Equation (4), which is known as Haldane's equation, expresses the relation between carboxyhemoglobin and arterial oxyhemoglobin concentrations in percentage (COHb%, $aO_2Hb\%$), and arterial CO and O_2 tensions (PaCO, PaO_2):

$$\frac{COHb\%}{aO_2Hb\%} = M\left(\frac{PaCO}{PaO_2}\right) \tag{4}$$

M represents affinity of CO for hemoglobin, relative to that of O_2. PaCO is calculated from ETCO (or alveolar CO concentration), P_{atm}, P_{vap} and arterial-alveolar CO tension difference (aAD_{CO}). Equation (4) is solved for COHb% by assigning an M value of 210 and the P_{atm} and P_{vap} values given above, as follows:

$$\begin{aligned} COHb\% &= 210 \times \frac{aO_2Hb\%}{PaO_2} \times PaCO \\ &= 210 \times \frac{aO_2Hb\%}{PaO_2} \times \left(94.7 \times ETCO + aAD_{CO}\right) \\ &= 210 \times \frac{aO_2Hb\%}{PaO_2} \times \left(1.04 \times PaCO_2 \times \frac{exCO}{exCO_2} + aAD_{CO}\right) \end{aligned} \tag{5}$$

PaO_2, $PaCO_2$ and aAD_{CO} may be considered as constant during the short time period required for cardiac output estimation (approximately 6 min). As CO and O_2 bind with hemoglobin competitively, sum of COHb% and $aO_2Hb\%$ is at a constant value between 90% and 100%. Thus, $aO_2Hb\%$ may also be considered as a constant, when fluctuation of COHb% is relatively small (below 5%). COHb% increase by infusion of 20 ml CO-saturated blood is approximately 0.3% to 0.6%, so equation (5) is simplified as follows:

$$COHb\% = K \times \frac{exCO}{exCO_2} + K' \tag{6}$$

K and K' are the constants peculiar to every estimation and subject, which is determined by the conditions of ventilation, oxygenation, and CO diffusion.

Determination of the Constants

The constants K and K' in equation (6) are determined by assigning two paired measurements of COHb% and $exCO/exCO_2$. For the first pair COHb% is that of venous blood sampled prior to the infusion measured using the hemoximeter, and $exCO/exCO_2$ is the average value during the baseline determination measured using the Carbolizer. For the second pair, COHb% is that of venous blood drawn 5 min 30 s after the infusion, and $exCO/exCO_2$ is the average value from 5 to 6 min after the infusion.

The time-COHb% curve by EGA is determined from continuous $exCO/exCO_2$ measurements using equation (6) and the determined values of K and K'.

Estimation of Cardiac Output and Circulating Blood Volume

Cardiac output by EGA is calculated from the time-COHb% curve and the infused volume of carboxyhemoglobin using the following equation (Stewart-Hamilton's method):

$$\text{Cardiac Output (liters/min)} = \frac{\text{Infused carboxyhemoglobin (g)}}{\text{AUC (g/l}\cdot\text{s)}} \times 60 \qquad (7)$$

The area under the curve (AUC), which represents the first pulmonary circulation of infused carboxyhemoglobin, is graphically determined from the time-COHb% curve by exponential regression and extrapolation of the early decay curve (` 1).

Circulating blood volume by EGA is calculated using the following equation:

$$\text{Circulating Blood Volume}\left(\text{liters}\right) = \frac{\text{Infused COHb\%} \times \text{infused volume (liters)}}{\text{Augmentation of COHb\%}} \times 100 \qquad (8)$$

The augmentation of COHb% by the infusion is calculated by exponential regression and extrapolation of the time-COHb% curve after the point major dilution phase was achieved.

Agreement of Cardiac Output Estimations between EGA and PAC

The objective of the experiment was to assess whether EGA can provide cardiac output estimation with accuracy equivalent to that of PAC. The subjects were 12 patients who were admitted to the intensive care unit attached to the Emergency Department of Saitama Medical Center. The patients required placement of PAC for diagnosis and treatment of circulatory disorders. Ten subjects required artificially controlled ventilation with endotracheal intubation and a respirator (Dr‰ger Medical AG & Co., EVITA4, Lubeck, Germany), and others required oxygen administration by mask. The patients or their legal guardians gave informed consent to the experiment.

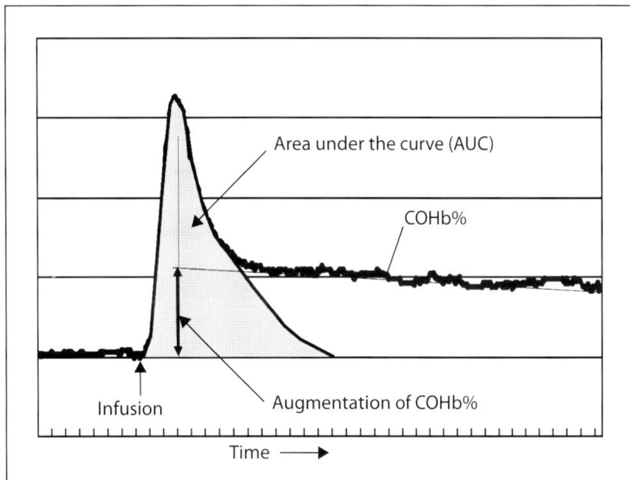

Fig. 1. The time-carboxyhemoglobin concentration (COHb%) curve in pulmonary circulation determined using carboxyhemoglobin densitometry by expiratory gas analysis. Cardiac output and circulating blood volume are estimated from the curve.

We estimated cardiac output by PAC using a Swan-Ganz standard thermodilution pulmonary artery catheter (131HF7, Edwards Lifesciences LLC, Irvine, Calif., USA). Following the bolus infusion of 5 ml ice-water via the right atrial lumen, cardiac output was computed from the time-temperature curve in pulmonary artery using a cardiac output computer (Edwards Lifesciences LLC, Vigilance Monitor). We compared an average of 3 consecutive estimations by PAC, at random respiratory phase, with a single estimation by EGA. We estimated cardiac outputs simultaneously by EGA (COEGA) and PAC (COPAC); twice in each subject with an interval of more than 12 h.

Differences between COEGA and COPAC are plotted against their means in figure 2. The distribution of the differences (mean ± SD) was –0.26 ± 0.25 liters/min. The 'limits of agreement' (mean +2 SD and mean –2 SD of the differences) were –0.75 liters/min and +0.23 liters/min. The 'coefficient of repeatability' (±2 SD of the deviations from the mean) of a single estimation by COPAC was ±1.00 liters/min or ±16% of the measurements.

Agreement of Circulating Blood Volumetry between EGA and the Conventional Method

The objective of the experiment was to assess whether EGA can provide circulating blood volumetry with accuracy equivalent to that of CO-hemoximetry, and with one-fifth infusion volume of CO-saturated blood. 18 adult healthy volunteers who gave written informed consent participated in the experiment.

First, circulating blood volume was estimated using EGA, with infusion of 20 ml of CO-saturated blood (CBVEGA). After a 3-hour interval, circulating blood volumetry

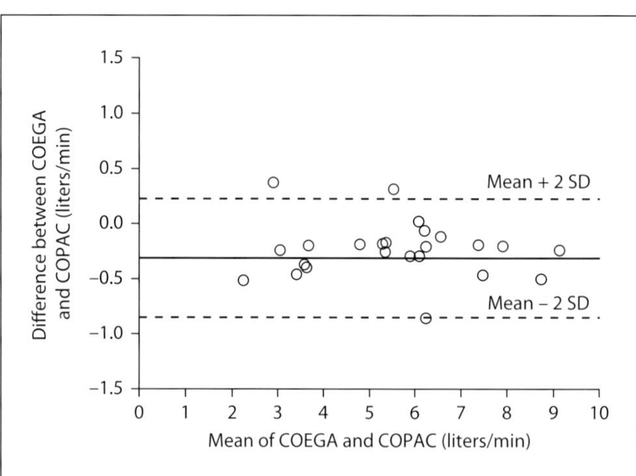

Fig. 2 Differences between cardiac output estimated by the EGA method (COEGA) and those by thermodilution using the pulmonary artery catheter (COPAC) are plotted against their means. The solid horizontal line indicates the mean of the differences, and the broken lines indicate the 'limits of agreement' (mean +2 SD and mean –2 SD of the differences).

was repeated using CO-hemoximetry with infusion of 100 ml of CO-saturated blood (CBVHEM). In CBVHEM, venous blood was sampled 10, 20 and 30 min after the infusion, to measure COHb% by CO-hemoximetry. Augmentation of COHb% by the infusion was calculated by exponential regression and extrapolation of these measurements.

Differences between CBVEGA and CBVHEM are plotted against their means in figure 3. The distribution of the differences among the 18 subjects was –0.02 ± 0.17 liters (mean ± SD). The 'limits of agreement' were –0.36 and 0.33 liters. Given the resolution of CO-hemoximetry (0.1%), the average resolution of the CBVHEM measurements was 0.33 liters.

Accuracy of Hemodynamic Monitoring by the EGA Method

The limits of agreement between cardiac output estimations by EGA and PAC (–0.75 and +0.23 liters/min) were included within the coefficient of repeatability for a single estimation by PAC (±1.00 liters/min or ±16% of the measurement). Reported coefficients of repeatability of PAC are 17–26% of the measurement for a single estimation, and 12–15% for an average of 3 repeated measurements [14–16]. Taken together, these findings indicate that the accuracy of EGA is equivalent to or greater than a single estimation by PAC, and is almost equivalent to the average of 3 repeated estimations by PAC if the constant bias between the methods is corrected. The bias between cardiac output estimations by EGA and PAC (–0.26 liters/min) indicates that PAC overestimates cardiac output compared to EGA. Figure 2 indicates that their relative differences (difference to mean ratios) are larger when the measurement is small. In PAC, a considerable amount of the infused tracer (heat) is lost from circulating blood

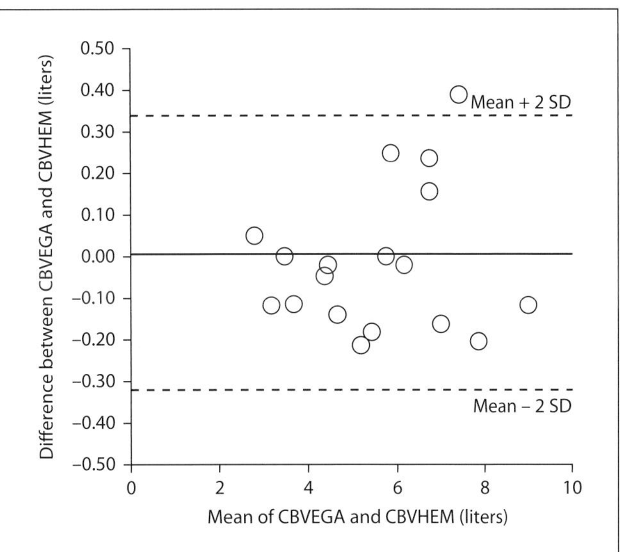

Fig. 3. Differences between circulating blood volume determined by the EGA method (CBVEGA) and those by the conventional method using hemoximetry (CBVHEM) are plotted against their means. The solid horizontal line indicates the mean of the differences, and the broken lines indicate the 'limits of agreement' (mean +2 SD and mean –2 SD of the differences).

before reaching the pulmonary artery when the flow is slow, resulting in overestimation of cardiac output in the lower range. In contrast, in EGA, no tracer (carboxyhemoglobin) infused into the peripheral vein is lost before reaching the pulmonary artery. This tracer loss accounts for the discrepancy in cardiac output estimation between the methods and suggests that EGA may be more accurate than PAC in the estimation of low cardiac output.

Given that the present average resolution of circulating blood volumetry using CO-hemoximetry was 0.33 liters, the limits of agreement between CBVEGA and CBVHEM indicate that application of EGA to carboxyhemoglobin dilution provided accuracy equivalent to that by CO-hemoximetry, with one-fifth the infusion volume.

Advantages of the EGA Method

Low invasiveness is the greatest advantage of EGA; the technique requires no pulmonary artery, central venous or arterial catheterization. EGA achieved accuracy almost equivalent to that of PAC by peripheral venous infusion of 20 ml of autologous blood saturated with CO. The infusion elevates COHb% by 0.3–0.6% in the average adult weighing 50–70 kg, which causes no significant adverse effects even in patients with anemia or hypoxia, or in heavy smokers [12]. Low cost is another advantage of EGA over PAC. The cost of EGA is approximately GBP 2 per estimation including CO gas and a disposable circuit for bubbling. Even if estimation is performed every 2 h for 3 days, the total cost per patient is relatively low compared with PAC, which costs over GBP 200 for a single-use thermodilution pulmonary artery catheter alone.

The continuous COHb% measurement following infusion of 50 ml of CO-saturated blood showed a transient increase exceeding 3.5% during the early mixing phase in some, and did not return to baseline within 30 min in most subjects. This indicates that the infusion of 100 ml of CO-saturated blood, which is required for carboxyhemoglobin dilution using CO-hemoximetry to achieve accuracy equivalent to the gold standard (^{51}Cr-labeled red cell dilution), induces an increase in COHb% greater than 7% and a further increase in the time required to return to baseline. The reported threshold of COHb% for clinical manifestation is approximately 10% in healthy adults, but is lower in patients with anemia, hypovolemia or cardiovascular disease. Therefore, the infusion volume must be significantly reduced to ensure safe, repeatable application of carboxyhemoglobin dilution, because hemodynamic monitoring is of primary importance in the management of such patients. Application of EGA to carboxyhemoglobin dilution provided equivalent accuracy with reduced infusion volume, and shortened the time required for elevated COHb% to return to baseline to enable repeated estimation with a shorter interval.

Disadvantages of the EGA Method

The disadvantage of EGA is the relatively long interval (approximately 1–3 h) between repeated estimations, which is required for elevated COHb% to return to baseline after the infusion. Consequently, EGA is incapable of continuous cardiac output estimation, which is already achieved by PAC or pulse contour analysis. Moreover, multiple estimations to improve the accuracy result in accumulation of carboxyhemoglobin and further prolongation of this interval. However, agreement between a single estimation by EGA and the average of three estimations by PAC in the present study suggests that a single estimation by EGA gives a clinically acceptable result, providing the constant bias between the methods is taken into account.

COHb% measurement by EGA is based on the assumption that alveolar gas and arterial blood have equivalent CO_2 tensions. As the diffusion capacity of CO_2 is very high (approximately 20 times that of oxygen), this assumption is valid for the most patients including those who are artificially ventilated. However, in patients with extreme ventilation-perfusion mismatch such as massive pulmonary embolism, the significant discrepancy between alveolar and arterial CO_2 tensions reduces the accuracy of COHb% measurement and hemodynamic monitoring by EGA.

Conclusions

Our study indicated that cardiac output estimation by low-dose carboxyhemoglobin dilution and EGA is acceptable as an equivalently accurate, minimally invasive and low-cost replacement of thermodilution using a PAC, providing the constant bias between the methods is taken into account. However, the long half-life of carboxyhemoglobin requires a relatively long interval between repeated estimations by EGA. EGA also achieved equivalently accurate circulating blood volumetry with one-fifth the infusion volume of the conventional method. EGA not only achieves safe and accurate hemodynamic monitoring in critically ill patients, but might also be useful for application in less severely ill patients, outpatients, or even athletes for the assessment of training effect.

References

1 Nilsson LB, Eldrup N, Berthelsen PG: Lack of agreement between thermo-dilution and carbon dioxide-rebreathing cardiac output. Acta Anaesthesiol Scand 2001;45:680–685.

2 Connors AF Jr, Speroff T, Dawson NV, et al: The effectiveness of right heart catheterization in the initial care of critically ill patients: SUPPORT investigators. JAMA 1996;276:889–897.

3 Harvey S, Harrison DA, Singer M, et al: Assessment of the clinical effectiveness of pulmonary artery catheters in management of patients in intensive care (PAC-Man): a randomised controlled trial. Lancet 2005;366:472–477.

4 Binanay C, Califf RM, Hasselblad V, et al: Evaluation study of congestive heart failure and pulmonary artery catheterization effectiveness: the ESCAPE trial. JAMA 2005;294:1625–1633.

5 Shah MR, Hasselblad V, Stevenson LW, et al: Impact of the pulmonary artery catheter in critically ill patients: meta-analysis of randomized clinical trials. JAMA 2005;294:1664–1670.

6 American Society of Anesthesiologists Task Force on Pulmonary Artery Catheterization: Practice guidelines for pulmonary artery catheterization: an updated report by the American Society of Anesthesiologists Task Force on Pulmonary Artery Catheterization. Anesthesiology 2003;99:988–1014.

7 International Committee for Standardization in Haematology: Standard techniques for the measurement of red cell and plasma volume. Br J Haematol 1973;25:801–814.

8 International Committee for Standardization in Haematology: Recommended methods for measurement of red cell and plasma volume. J Nucl Med 1980;21:793–800.

9 Fukui M, Shigemi K: Determination of single and repeated red cell volumes by the indicator dilution method using carbon monoxide as the indicator. Crit Care Med 1989;17:1199–1202.

10 Obata H, Goto F, Nara T, et al: High predictive value of red cell volume measurement using carboxyhaemoglobin in a rabbit model of haemorrhage. Br J Anaesth 1998;81:940–944.

11 Ohki S, Kunimoto F, Isa Y, et al: Accuracy of carboxyhemoglobin dilution method for the measurement of circulating blood volume. Can J Anaesth 2000;47:150–154.

12 Sawano M, Mato T, Tsutsumi H: Bedside red cell volumetry by low-dose carboxyhaemoglobin dilution using expiratory gas analysis. Br J Anaesth 2006;96:186–194.

13 Bland JM, Altman DG: Statistical methods for assessing agreement between two methods of clinical measurement. Lancet 1986;1:307–310.

14 Kohanna FH, Cunningham JN Jr: Monitoring of cardiac output by thermo-dilution after open-heart surgery. J Thorac Cardiovasc Surg 1977;73:451–457.
15 Stetz CW, Miller RG, Kelly GE, Raffin TA: Repeatability of PAC method in the determination of cardiac output in clinical practice. Am Rev Respir Dis 1982;126:1001–1004.
16 Staiar K, Wiesenack C, Günkel L, Keyl C: Cardiac output determination by thermo-dilution and arterial pulse waveform analysis undergoing aortic valve replacement. Can J Anaesth 2008;55:22–28.

Makoto Sawano
Department of Emergency Medicine and Critical Care, Saitama Medical Centre, Saitama Medical University
1981 Kamoda, Kawagoe
Saitama, 350-8550 (Japan)
Tel./Fax +81 49 228 3755, E-Mail sawanom-tky@umin.ac.jp

Yoshikawa T, Naito Y (eds): Gas Biology Research in Clinical Practice.
Basel, Karger, 2011, pp 35–42

CO and Its Application to Gastrointestinal Disease

Yuji Naito · Kazuhiko Uchiyama · Tomohisa Takagi · Toshikazu Yoshikawa

Molecular Gastroenterology and Hepatology, Graduate School of Medical Science, Kyoto Prefectural University of Medicine, Kyoto, Japan

Abstract

Carbon monoxide (CO) has been reported to be a biologically active product. CO is an activator of soluble guanylate cyclase (GC) and involved in increased cGMP production, activation of type I cGMP-dependent protein kinase and in smooth muscle relaxation. CO also modulates cGMP levels and leads to cGMP-dependent protein kinase I. This is one of the targets of CO that acts as a smooth muscle relaxant by direct effects on the contractile machinery as well as by altering Ca^{2+} homeostasis and voltage-gated ion channel activity. The K^+ channel, amongst others, is activated through the CO-dependent increase of cGMP. In the body, two isoforms of heme oxygenases (HO-1, HO-2) catalyze the synthesis of CO from Fe protoporphyrin IX (heme). Concerning the pathogenesis of inflammatory disease of the gastrointestinal tract, the HO-1 level is elevated mainly in infiltrative inflammatory cells in the intestinal mucosa in patients with active ulcerative colitis. Not only endogenous CO but also exogenous CO has been reported to have anti-inflammatory and anti-apoptotic effects by various mechanisms and are involved in attenuating colonic mucosal inflammation. Recently, CO has been reported to have other functions such as colonic epithelial restitution. Further study is needed to reveal a beneficial role of CO in the clinical field.

Carbon monoxide (CO) has been reported as a biologically active product of heme metabolism by heme oxygenase (HO)-1 that regulates neurotransmission, smooth muscle tone and the response to cellular injury. Biliverdin is also a product of heme metabolism by HO-1 and these byproducts have been reported to include cellular protective effects such as anti-apoptotic and anti-oxidant actions against oxidative stress [1–4]. In addition to environmental exposure, a considerable amount of CO arises endogenously as a product of ordinary metabolism. The majority of blood CO-Hb arises from endogenous production, corresponding to blood CO levels of 0.4–1% [5]. These values increase not only with the environmental background, but also under pathological or toxicological conditions that produce global or tissue-specific elevations of HO-1.

The role of CO in the gastrointestinal (GI) tract has not been clarified; however, the field has advanced recently, with clear demonstrations of CO as a nerve-derived signaling molecule and as a suppressor of cellular injury. In this review, we will discuss the role of CO as a suppressor of GI disease.

Physiological Role of CO in the Gastrointestinal Tract

In 1991, the role of endogenous CO, its metabolic reason and a physiological meaning were suggested [6]. The role of CO as an endogenous neural messenger, based on the effect of HO inhibitors and the histological location of HO was detected in 1993 [7]. At the same time, it was reported that endogenous CO was also involved in the relaxation of the opossum internal anal sphincter (IAS) in response to nonadrenergic noncholinergic (NANC) nerve stimulation [8].

The role and distribution of HO-2 was first detected in the GI tract. HO-2 is a constitutive HO and it is important as a CO provider under physiological conditions. In addition to the neuronal expression of HO-2 in the submucosal and myenteric plexus of the GI tract, its immunoreactivity was also discovered in non-neuronal cells such as smooth muscle and arterial endothelium [9]. Several studies demonstrated CO from HO-2 as a NANC and CO is the major mediator for relaxation of the intestine [10].

Mechanism of CO Action

CO is known to be an activator of soluble guanylate cyclase (GC). Though CO is a weak activator of GC in vitro with much lower potency and efficacy than NO, the application of CO to a number of different tissues results in increased cGMP production, activation of type I cGMP-dependent protein kinase and smooth muscle relaxation [11, 12], suggesting that CO does modulate cGMP levels in vivo. Several mechanisms have been proposed to explain the enhanced effects of CO on soluble GC in vivo. One suggests that the synthetic molecule YC-1 acts in synergism with CO as an activator of soluble GC [13], another suggests that NO and CO act together to increase cGMP production [14]. However, it is not yet clear. The activation of cGMP-dependent protein kinase I is one of the targets of CO that acts as a smooth muscle relaxant by its direct effects on the contractile machinery as well as by altering Ca^{2+} homeostasis and voltage-gated ion channel activity [15]. CO has also reported to activate K^+ channels in a variety of tissues, including the GI tract. Intracellular cGMP activates the K^+ channel and the cGMP level is increased by the treatment of exogenous CO. As exogenous CO does not affect the K^+ channel, CO activates the K^+ channel through cGMP [16].

CO has been reported to activate K^+ channels in a variety of tissues, including the GI tract. In human intestinal smooth muscle, CO induces membrane hyperpolarization by a delayed rectifier-like K^+ current [17, 18].

The anti-apoptotic potential of CO has been reported. TNF-α-induced apoptosis in mouse fibroblasts [19] and endothelial cells [20] was inhibited by exogenous CO treatment. This anti-apoptotic effect of CO is reported to depend on the p38 MAPK pathway [20]. On the other hand, in Jurkatt T cells, CO treatment increased Fas/CD95-induced apoptosis.

Distribution and Regulation of Heme Oxygenase in the Gastrointestinal Tract

Two isoforms of heme oxygenases (HO-1, HO-2) catalyze the synthesis of CO from Fe protoporphyrin IX (heme) [11]. Though these proteins are structurally quite different, the reaction chemistry is the same, relying on oxidation of NADPH and using molecular oxygen in the cleavage of heme. CO, biliverdin, Fe^{2+} and H_2O_2 are generated by this reaction. In the normal condition, HO-1 is expressed at very low levels in the GI tract. It is induced by disease, injury and/or inflammation. On the other hand, HO-2 has been reported to express in neuronal cell bodies and fibers in the deep muscular plexuses [21]. Neurons in the pyloric and ileocecal sphincters are reported to have particularly high levels of HO-2 [22]. Non-neuronal cells in the GI tract also express HO-2, including a population of cells in the mucosal epithelium, the smooth muscle of blood vessels, the endothelium of blood vessels, and the interstitial cells of Cajal [18, 23–26]. Regarding HO-1, our previous report has shown that mRNA and protein expression are significantly elevated in mainly infiltrative inflammatory cells in the intestinal mucosa in patients with active ulcerative colitis, and upregulated HO-1 might contribute to an anti-inflammatory effect [27].

Anti-Apoptotic and Anti-Inflammatory Effect of CO

The anti-apoptotic effect of CO was first reported in an in vitro model using a TNF-α-initiated apoptosis model. Brouard et al. [20] reported that CO inhibited TNF-α-initiated apoptosis in mouse endothelial cells through activation of the p38 MAPK pathway. Furthermore, HO-1 or CO cooperated with NF-κB-dependent anti-apoptotic genes to protect against TNF-α-mediated endothelial cell apoptosis [28]. Low-dose CO pretreatment has been reported to show an anti-apoptotic effect in several models of disease or tissue injury. Exogenously applied CO inhibited ischemia/reperfusion-induced apoptosis associated with the CO-dependent activation of the p38 MAPK and its upstream MAPK kinase (MKK3), and with the suppression of ERK and JNK activation [29].

Anti-inflammatory effect of CO has been reported using cell culture and animal models of sepsis [30]. Lipopolysaccharide (LPS) stimulation induces proinflammatory cytokines (i.e. TNF-α) in macrophages (RAW 264.7). CO administration or HO-1 overexpression in RAW 264.7 cells inhibit TNF-α expression treated with LPS. In one of the mechanisms, it has been reported that CO treatment inhibited the LPS-induced activation of NF-kB by preventing phosphorylation and degradation of the inhibitory subunit I-kBa, and this mechanism was associated with granulocyte-macrophage colony-stimulating factor (GM-CSF) modulation [31]. GM-CSF is a glycoprotein which has been reported to enhance the secretion of proinflammatory cytokines including TNF-α, IL-1 and IFN-γ [32]. In several inflammatory models, CO has been reported to inhibit GM-CSF expression resulting in attenuation of inflammation.

Inflammatory Bowel Disease and CO

The pathogenesis of inflammatory bowel disease (IBD) such as Crohn's disease (CD) and ulcerative colitis (UC) is complex and still unclear. There are several reports which suggest cigarette smoking shows a protective effect against the development of UC [33, 34]. The detailed mechanism is unclear, but CO, one of the component of cigarette smoking, has been reported to ameliorate colonic inflammation.

The effect of CO to improve colonic inflammation has been reported in mouse colitis models. The IL-10-deficient (*IL-10*$^{-/-}$) mouse develops T helper (Th)1-mediated colitis, and it is used as a model for chronic colitis. CO was exposed at a concentration of 250 ppm for 7 days and CO exposure ameliorates the colitis in the *IL-10*$^{-/-}$ mouse [35]. It has reported that CO alters IFN-γ signaling in macrophages, and decreased IFN regulatory factor (IRF)-8 and IL-12 p40 expression. It was found that CO antagonizes the inhibitory effect of IFN-γ on HO-1 expression in macrophages.

CO inhalation is reported to ameliorate 2,4,6-trinitrobenzine sulfonic acid (TNBS)-induced mice colitis [36]. Macroscopic colonic damage score, thiobarbituric acid-reactive substances, and tissue-associated myeloperoxidase (MPO) activity in colonic mucosa were reduced by CO treatment (fig. 1). Colonic mucosal TNF-α expression and TNF-α production by CD4+ T cells isolated from the spleen were significantly inhibited. In that study, the beneficial effects of CO in a murine colitis model may be attributed to its anti-inflammatory properties.

Not only reducing mucosal inflammation, but also colonic mucosal healing is one of the important factors to control the pathogenesis of IBD. The effect of CO to colonic epithelial cell restitution has been reported [37]. It has been suggested that submucosal myofibroblasts play a very important role in epithelial cell restitution via transforming growth factor (TGF)-β secretion. But CO induces fibroblast growth factor 15 (FGF15) expression in mouse colonic myofibroblasts via the inhibition of mir710, and FGF15 enhances the restitution of mouse colonic epithelial cells (fig. 2) [37].

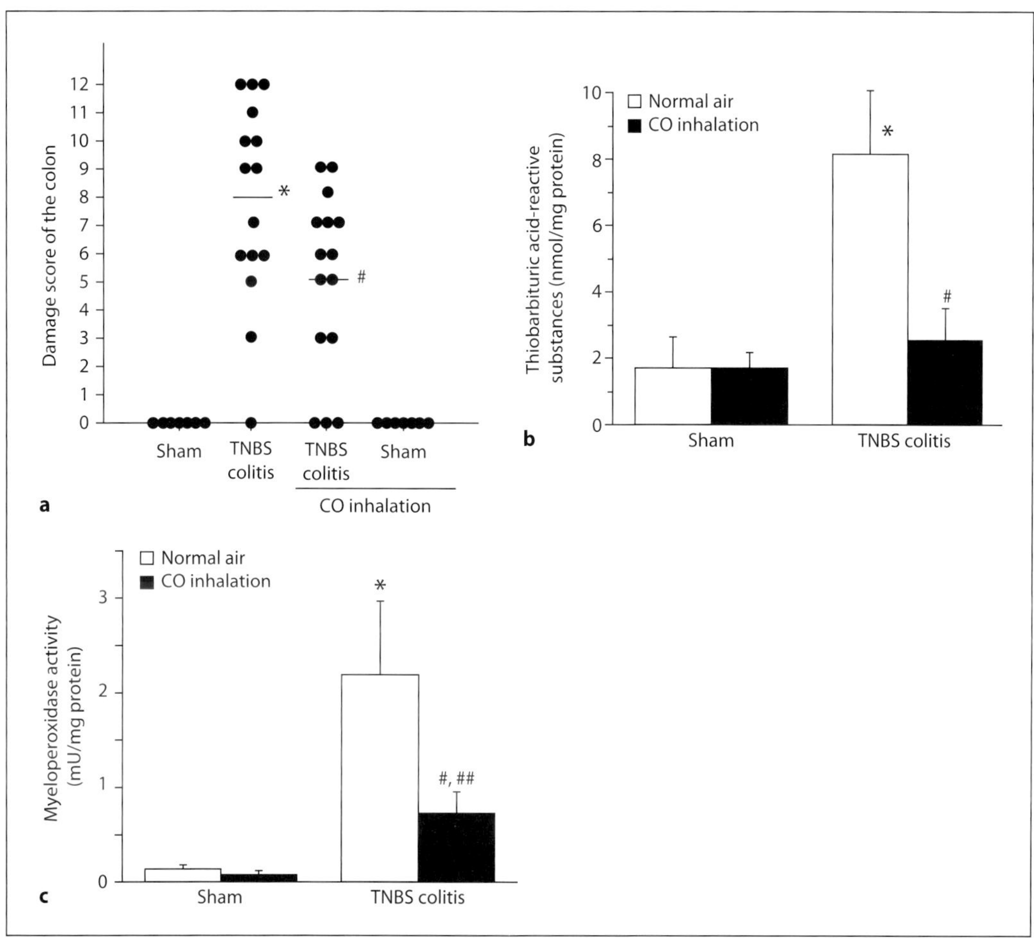

Fig. 1. a Damage score in a TNBS colitis model treated with CO inhalation (* $p < 0.01$, # $p < 0.05$ vs. sham). **b** TBA level of colonic musosa in a TNBS colitis model treated with CO inhalation. * $p < 0.01$, # $p < 0.05$ vs. sham. **c** MPO activity of colonic mucosa in a TNBS colitis model treated with CO inhalation. * $p < 0.01$, # $p < 0.05$ vs. sham [36].

These findings suggest that CO has the possibility of being a therapeutic agent for IBD because of its anti-inflammatory effects and colonic mucosal healing. Though further studies are needed to provide new effects of CO in IBD, the beneficial effect of CO to IBD is getting reported nowadays.

Conclusions

The physiological role of CO and its acute as well as chronic regulation in the GI tract has not been fully addressed. Recently, the anti-inflammatory effect of CO has been

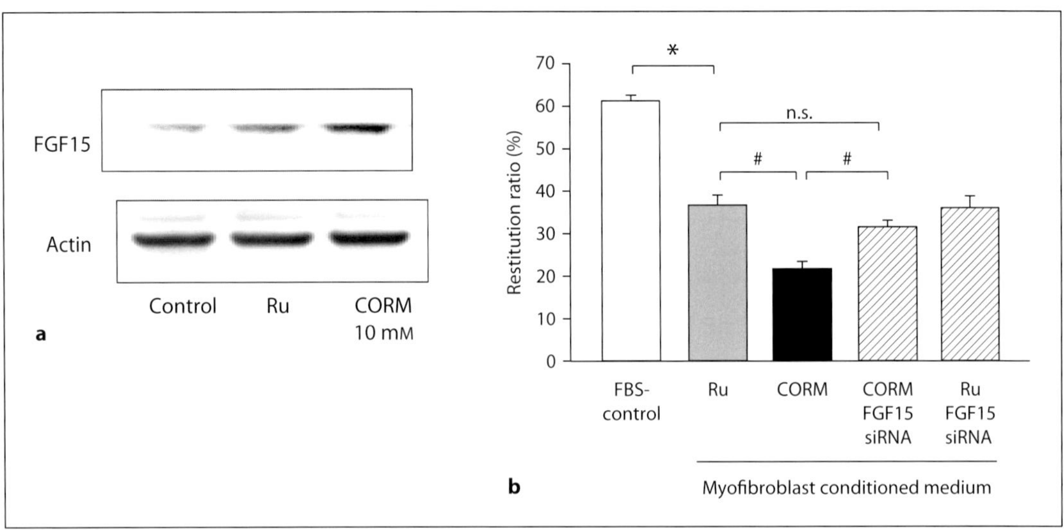

Fig. 2. a CORM (CO-releasing molecule)-induced FGF15 protein expression in colonic myofibroblasts. Ru = Ruthenium, control of CORM. **b** FGF15 is related to colonic epithelial cell restitution. * $p < 0.01$, # $p < 0.05$ [37].

focused on as a therapeutic agent in the GI tract, especially the control of IBD. CO treatment in an experimental murine colitis model and murine colonic epithelial and stromal cells were discussed and the multiple function of CO has now been revealed, not only as an anti-inflammatory but also for epithelial cellular restitution in the GI tract.

References

1 Alcaraz MJ, Fernandez P, Guillen MI: Anti-inflammatory actions of the heme oxygenase-1 pathway. Curr Pharm Des 2003;9:2541–2551.

2 Lee TS, Chau LY: Heme oxygenase-1 mediates the anti-inflammatory effect of interleukin-10 in mice. Nat Med 2002;8:240–246.

3 Morse D, Choi AM: Heme oxygenase-1: from bench to bedside. Am J Respir Crit Care Med 2005;172:660–670.

4 Nakao A, Kaczorowski DJ, Sugimoto R, Billiar TR, McCurry KR: Application of heme oxygenase-1, carbon monoxide and biliverdin for the prevention of intestinal ischemia/reperfusion injury. J Clin Biochem Nutr 2008;42:78–88.

5 Von Burg R: Carbon monoxide. J Appl Toxicol 1999; 19:379–386.

6 Marks GS, Brien JF, Nakatsu K, McLaughlin BE: Does carbon monoxide have a physiological function? Trends Pharmacol Sci 1991;12:185–188.

7 Verma A, Hirsch DJ, Glatt CE, Ronnett GV, Snyder SH: Carbon monoxide: a putative neural messenger. Science 1993;259:381–384.

8 Rattan S, Chakder S: Inhibitory effect of CO on internal anal sphincter: heme oxygenase inhibitor inhibits NANC relaxation. Am J Physiol 1993;265: G799–804.

9 Zakhary R, Gaine SP, Dinerman JL, Ruat M, Flavahan NA, Snyder SH: Heme oxygenase 2: endothelial and neuronal localization and role in endothelium-dependent relaxation. Proc Natl Acad Sci USA 1996;93:795–798.

10 Watkins CC, Boehning D, Kaplin AI, Rao M, Ferris CD, Snyder SH: Carbon monoxide mediates vasoactive intestinal polypeptide-associated nonadrenergic/noncholinergic neurotransmission. Proc Natl Acad Sci USA 2004;101:2631–2635.
11 Maines MD: The heme oxygenase system: a regulator of second messenger gases. Annu Rev Pharmacol Toxicol 1997;37:517–554.
12 Denninger JW, Marletta MA: Guanylate cyclase and the NO/cGMP signaling pathway. Biochim Biophys Acta 1999;1411:334–350.
13 Friebe A, Schultz G, Koesling D: Sensitizing soluble guanylyl cyclase to become a highly CO-sensitive enzyme. EMBO J 1996;15:6863–6838.
14 Thorup C, Jones CL, Gross SS, Moore LC, Goligorsky MS: Carbon monoxide induces vasodilation and nitric oxide release but suppresses endothelial NOS. Am J Physiol 1999;277:F882–F889.
15 Carvajal JA, Germain AM, Huidobro-Toro JP, Weiner CP: Molecular mechanism of cGMP-mediated smooth muscle relaxation. J Cell Physiol 2000;184:409–420.
16 Liu H, Mount DB, Nasjletti A, Wang W: Carbon monoxide stimulates the apical 70-pS K^+ channel of the rat thick ascending limb. J Clin Invest 1999;103: 963–970.
17 Farrugia G, Irons WA, Rae JL, Sarr MG, Szurszewski JH: Activation of whole cell currents in isolated human jejunal circular smooth muscle cells by carbon monoxide. Am J Physiol 1993;264:G1184–1189.
18 Farrugia G, Miller SM, Rich A, Liu X, Maines MD, Rae JL, Szurszewski JH: Distribution of heme oxygenase and effects of exogenous carbon monoxide in canine jejunum. Am J Physiol 1998;274: G350–G358.
19 Petrache I, Otterbein LE, Alam J, Wiegand GW, Choi AM: Heme oxygenase-1 inhibits TNF-alpha-induced apoptosis in cultured fibroblasts. Am J Physiol Lung Cell Mol Physiol 2000;278:L312–L319.
20 Brouard S, Otterbein LE, Anrather J, Tobiasch E, Bach FH, Choi AM, Soares MP: Carbon monoxide generated by heme oxygenase 1 suppresses endothelial cell apoptosis. J Exp Med 2000;192:1015–1026.
21 Miller SM, Reed D, Sarr MG, Farrugia G, Szurszewski JH: Haem oxygenase in enteric nervous system of human stomach and jejunum and co-localization with nitric oxide synthase. Neurogastroenterol Motil 2001;13:121–131.
22 Ny L, Alm P, Larsson B, Andersson KE: Morphological relations between haem oxygenases, NO-synthase and VIP in the canine and feline gastrointestinal tracts. J Auton Nerv Syst 1997;65:49–56.
23 Grozdanovic Z, Gossrau R: Expression of heme oxygenase-2 (HO-2)-like immunoreactivity in rat tissues. Acta Histochem 1996;98:203–214.
24 Ny L, Alm P, Ekstrom P, Larsson B, Grundemar L, Andersson KE: Localization and activity of haem oxygenase and functional effects of carbon monoxide in the feline lower oesophageal sphincter. Br J Pharmacol 1996;118:392–399.
25 Zakhary R, Poss KD, Jaffrey SR, Ferris CD, Tonegawa S, Snyder SH: Targeted gene deletion of heme oxygenase 2 reveals neural role for carbon monoxide. Proc Natl Acad Sci USA 1997;94:14848–14853.
26 Hu Y, Yang M, Ma N, Shinohara H, Semba R: Contribution of carbon monoxide-producing cells in the gastric mucosa of rat and monkey. Histochem Cell Biol 1998;109:369–373.
27 Takagi T, Naito Y, Mizushima K, Nukigi Y, Okada H, Suzuki T, Hirata I, Omatsu T, Okayama T, Handa O, Kokura S, Ichikawa H, Yoshikawa T: Increased intestinal expression of heme oxygenase-1 and its localization in patients with ulcerative colitis. J Gastroenterol Hepatol 2008;23(suppl 2):S229–S233.
28 Brouard S, Berberat PO, Tobiasch E, Seldon MP, Bach FH, Soares MP: Heme oxygenase-1-derived carbon monoxide requires the activation of transcription factor NF-kappa B to protect endothelial cells from tumor necrosis factor-alpha-mediated apoptosis. J Biol Chem 2002;277:17950–17961.
29 Zhang X, Shan P, Otterbein LE, Alam J, Flavell RA, Davis RJ, Choi AM, Lee PJ: Carbon monoxide inhibition of apoptosis during ischemia-reperfusion lung injury is dependent on the p38 mitogen-activated protein kinase pathway and involves caspase 3. J Biol Chem 2003;278:1248–1258.
30 Otterbein LE, Bach FH, Alam J, Soares M, Tao Lu H, Wysk M, Davis RJ, Flavell RA, Choi AM: Carbon monoxide has anti-inflammatory effects involving the mitogen-activated protein kinase pathway. Nat Med 2000;6:422–428.
31 Sarady JK, Otterbein SL, Liu F, Otterbein LE, Choi AM: Carbon monoxide modulates endotoxin-induced production of granulocyte macrophage colony-stimulating factor in macrophages. Am J Respir Cell Mol Biol 2002;27:739–745.
32 Reed JA, Whitsett JA: Granulocyte-macrophage colony-stimulating factor and pulmonary surfactant homeostasis. Proc Assoc Am Physicians 1998;110: 321–332.
33 Jick H, Walker AM: Cigarette smoking and ulcerative colitis. N Engl J Med 1983;308:261–263.
34 Boyko EJ, Koepsell TD, Perera DR, Inui TS: Risk of ulcerative colitis among former and current cigarette smokers. N Engl J Med 1987;316:707–710.

35 Hegazi RA, Rao KN, Mayle A, Sepulveda AR, Otterbein LE, Plevy SE: Carbon monoxide ameliorates chronic murine colitis through a heme oxygenase 1-dependent pathway. J Exp Med 2005;202:1703–1713.

36 Takagi T, Naito Y, Mizushima K, Akagiri S, Suzuki T, Hirata I, Omatsu T, Handa O, Kokura S, Ichikawa H, Yoshikawa T: Inhalation of carbon monoxide ameliorates TNBS-induced colitis in mice through the inhibition of TNF-alpha expression. Dig Dis Sci 2010;55:2797–2804.

37 Uchiyama K, Naito Y, Takagi T, Mizushima K, Hayashi N, Harusato A, Hirata I, Omatsu T, Handa O, Ishikawa T, Yagi N, Kokura S, Yoshikawa T: Carbon monoxide enhance colonic epithelial restitution via FGF15 derived from colonic myofibroblasts. Biochem Biophys Res Commun 2010;391:1122–1126.

Dr. Yuji Naito
Molecular Gastroenterology and Hepatology, Graduate School of Medical Science
Kyoto Prefectural University of Medicine
465 Kajiicho Hirokoji Kawaramachi-dori, Kamigyo-ku
Kyoto 602-8566 (Japan)
Tel. +81 75 251 5518, Fax +81 75 251 0710, E-Mail ynaito@koto.kpu-m.ac.jp

Yoshikawa T, Naito Y (eds): Gas Biology Research in Clinical Practice.
Basel, Karger, 2011, pp 43–55

Clinical Application of Inhaled Nitric Oxide

Kazuo Maruyama[a] · Erquan Zhang[a] · Junko Maruyama[b]

[a]Department of Anesthesiology and Critical Care Medicine, Mie University School of Medicine, and
[b]Department of Clinical Engineering, Suzuka University of Medical Science, Mie, Japan

Abstract

Inhaled nitric oxide (NO) is a selective pulmonary vasodilator because it does not reduce systemic blood pressure. Inhaled NO dilates constricted pulmonary vessels by activating guanylate cyclase of pulmonary vascular smooth muscle cells and is inactivated by hemoglobin when it diffuses into the bloodstream. Inhaled NO improves arterial oxygenation in the lungs with increased intrapulmonary shunt by shifting the blood flow to ventilated units accessible to NO from nonventilated units. In the lung with a broad distribution of units with low ventilation perfusion (V/Q) ratios, inhaled NO alone reduces arterial oxygenaton by deteriorating the V/Q mismatch through inhibition of hypoxic pulmonary vasoconstriction, so combined inhalation of oxygen is necessary to achieve NO-induced improvement of oxygenation. Inhaled NO is a registered treatment for full-term infants with persistent pulmonary hypertension of neonates and might be indicated in pulmonary vasoreactivity testing for cardiac catheterization, patients with cardiac surgery, treatment of acute right-sided heart failure in cardiac transplant recipients, left ventricular assist device patients with elevated pulmonary vascular resistance, and high-altitude pulmonary edema. A meta-analysis of 12 randomized controlled trials does not indicate the routine use of inhaled NO in patients with acute respiratory distress syndrome (ARDS), although it is currently performed as a rescue therapy in patients with severe ARDS who require extracorporeal membrane oxygenation. Inhaled NO might reduce reperfusion injury and inflammation in remote organs other than the lung such as for liver transplantation, which is mediated by NO containing intermediates such as nitrite and nitrosothiol and could become a new indication of inhaled NO.

Nitric oxide (NO) as a physiological substance was first discovered as an endothelium derived-relaxing factor (EDRF) and is synthesized by nitric oxide synthase (NOS) when L-arginine is converted to L-citrulline. The reaction needs oxygen, cofactors (thiol, flavin, biopterin) and nicotinamide adenine dinucleotide phosphate (NADPH) [1–4]. NO reacts with oxygen, transitional metal ions, thiol, and superoxide [4]. The major physiological effects of nitric oxide include (1) relaxation of vascular smooth muscle cells, (2) inhibition of leukocyte adhesion, (3) inhibition of platelet adhesion and activation, and (4) inhibition of cell proliferation [5–8]. NO activates guanylate cyclase by

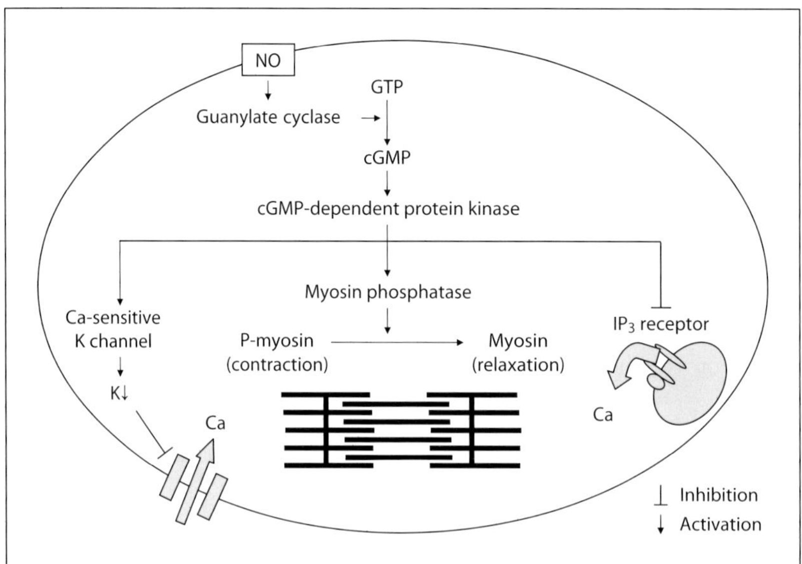

Fig. 1. Activation of guanylate cyclase. Activation of guanylate cyclase → increase in cyclic guanosine monophosphate (cGMP) → activation of cGMP dependent protein kinase → decrease in Ca concentration in the cytoplasm and decrease in Ca sensitivity of myosin → relaxation. The decrease in Ca concentration in the cytoplasm is induced by: (1) the activation of Ca sentivive potassium channel → decrease in membrane potential → inhibition of L-type calcium channel → decrease in calcium influx into the cytoplasm, and (2) phosphorylation of the protein (IRAG)-forming complex with inositol triphosphate → inhibition of calcium release from intracellular calcium store sites, sarcoplasmic reticulm → decrease in calcium concentration in the cytoplasm. The decrease in Ca sensitivity of myosin is induced by: phosphorylation of myosin-binding subunit of myosin phosphatase → activation of myosin light-chain phosphatase → dephosphorylation of phosphorylated myosin → decrease in contraction at the same concentration of calcium, which means a decrease in Ca sensitivity of myosin.

reacting with Fe in guanylate cyclase and induces pulmonary vascular smooth muscle cell relaxation as described in figure 1 [7]. When NO diffuses into the bloodstream, it binds with the components (Fe, heme transitional metal) of hemoglobin (Hb) and is inactivated. The greater part of the metabolites of inhaled NO is excreted in the urine in the form of nitrates (NO_3^-) [9]. Because NO is a gas, it can be delivered to the lung by inhalation and works mainly in the lung for its metabolic fate. Although inhaled NO dilates the pulmonary vasculature, it does not induce systemic vasodilation [10–14]. Thus, inhaled NO has a unique position as a selective pulmonary vasodilator. NO inhalation represents a practical application of basic animal research.

Impairment of NO-Dependent Responses in Pulmonary Hypertension

The endothelium-dependent relaxation in response to acetylcholine is impaired in isolated pulmonary arteries from humans with end-stage chronic obstructive pulmonary

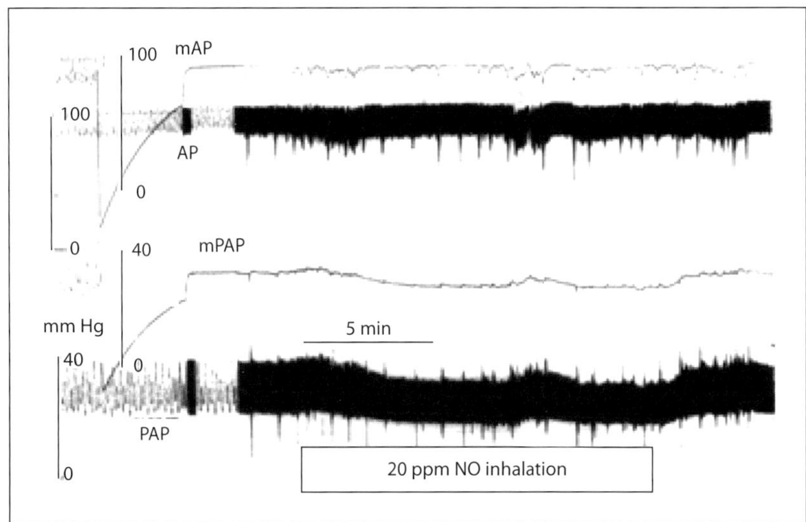

Fig. 2. Selective pulmonary vasodilation in pulmonary hypertension in rats. 20 ppm NO inhalation selectively reduced pulmonary artery pressure without changing systemic artery pressure in pulmonary hypertensive rats with pulmonary vascular remodeling. Indwelling pulmonary and artery catheters were placed and the pressures were recorded with rat fully conscious [48].

disease (COPD) [15] and rats exposed to chronic hypoxia [16–18]. These impairments might be due to (1) reduction of NO release from injured pulmonary vascular endothelial cells, (2) reduction of vasodilation responses to NO in hypertensive pulmonary vascular smooth muscle cells, or (3) impairment of diffusion from endothelial cells to smooth muscle cells by scavenging or distance between these cells. Production of NO during hypoxic exposure increases in the whole body because the plasma nitrate concentration is significantly increased [19]. Endothelial nitric oxide synthase (eNOS)-specific immunoreactivity was increased in peripheral pulmonary arteries and eNOS protein levels in whole lung tissue were increased in chronic hypoxia-induced pulmonary hypertension in rats [19], but not in iNOS [20]. Thus, production of NO in the lung might not be reduced in chronic hypoxia-induced PH, suggesting that the increase in NO by eNOS might be a compensatory mechanism and inhalation is regarded as a supplementation therapy. Regarding the effect of NO on the smooth muscles located beneath the endothelium, the relaxation response to NO donor sodium nitroprusside (SNP) is depressed in isolated pulmonary arteries from hypoxia-induced PH rats [16, 18], suggesting a decreased sensitivity of smooth muscle to NO. Clinically, the important point is that although the response to NO is depressed compared to normal control pulmonary arteries, the relaxation response to NO is still present in the structurally remodeled pulmonary artery [21, 22]. Because inhaled NO induces selective pulmonary vasodilatation without systemic vasodilation [10–14], this therapy would be of benefit for patients with hypertensive pulmonary vascular disease (fig. 2).

Discovery of NO Inhalation Therapy in Animals and Humans

Elective pulmonary vascular relaxation induced by inhaled NO was first described in animal models [10] and in patients with pulmonary hypertension (PH) [11]. Inhaling 80 parts per million (ppm) NO did not change the normal pulmonary artery pressure (PAP), but decreased heightened PAP with hypoxic pulmonary vasoconstriction or U46619 (thromboxane analogue) induced pulmonary vasoconstriction [10]. Thus, inhaled NO dilates constricted pulmonary vasculature in animals. Since the pulmonary vasculature is maximally dilated in the unstimulated condition, inhaled NO has little opportunity to reverse the constriction in this situation. In patients with PH, 40 ppm NO inhalation decreased the pulmonary vascular resistance (PVR) by 5–68% from baseline without changes in systemic vascular resistance (SVR). In contrast, prostacyclin (24 μg/h) infusion induced the reductions in both PVR and SVR. The magnitude of the reduction in PVR induced by 40 ppm NO inhalation was similar to the reduction induced by prostacyclin (24 μg/h). In normal volunteers, 40 ppm inhaled NO induced no changes in either PVR or SVR. Thus, in patients with pulmonary hypertension, inhaled NO induces selective pulmonary vasodilation unlike prostacyclin infusion which causes both pulmonary and systemic vasodilation [11].

Physiological Effect of NO Inhalation on Circulation

Pulmonary vasodilation results in a decrease in pulmonary artery pressure, decrease in PVR, reduction of right ventricular afterload, decrease in intrapulmonary shunt, changes in distribution of ventilation/perfusion ratio, and decrease in venous admixture. These effects might have be beneficial for in treating PH [23–25], right ventricular failure [35, 36], hypoxemic respiratory failure [26, 27, 31–34, 37] and pulmonary gas exchange [28–30, 38].

Effects of Inhaled NO on the PAP

In the first studies of NO inhalation as therapeutic purpose in animal and human [10, 11], the inhaled NO concentrations were 5–80 and 40 ppm, respectively. In an animal model, vasoconstriction was induced by hypoxic exposure and thromboxane analogue infusion, in which the pulmonary vascular bed was normal without hypertensive vascular remodeling. 5 ppm NO inhalation significantly reduced PAP and an almost complete vasodilator response occurred by 40 or 80 ppm NO [10] in constricted pulmonary arteries. In contrast, the patients had an abnormal pulmonary vascular bed with hypertensive pulmonary vascular remodeling. NO inhalation does not change the PAP in the unstimulated normal lung, but decreases the PAP in the hypertensive lung probably because the structurally remodeled pulmonary vasculature has a heightened vascular reactivity and has tone in the unstimulated state. We determined the dose-response relationship between inhaled NO concentration and the reduction of PAP in monocrotaline-induced PH rats having an indwelling pulmonary artery catheter with the rats fully awake [21]. NO of 20–100 ppm was inhaled and we found that PAP was reduced dose-dependently at concentrations of 20–60 ppm and the response to NO became constant at concentrations above 60 ppm. In a later rat study, no apparent dose-dependent effect of NO on mean PAP was detected for 5, 10, 20 and 40 ppm NO inhalation in monocrotaline-induced PH [22].

Low Dose NO Inhalation on PAP

The relationship between the hypertensive vascular changes and response to low-dose inhaled NO was determined in chronic hypoxia-induced PH in rats [39]. Inhaled NO decreased PAP in chronic hypoxia-induced PH in rats from the concentrations as low as 0.1 ppm [39]. The magnitudes of the reduction were similar between 0.1 and 2.0 ppm. The absolute value of the decrease in mean PAP with NO inhalation, but not percent reduction in mean PAP, was significantly correlated with baseline mean PAP and the severity of pulmonary vascular changes, i.e. new muscularization of normally nonmuscular peripheral pulmonary arteries and medial hypertrophy of muscular arteries [39]. This indicates that in chronic hypoxia-induced PH the severity of pulmonary remodeling does not alter the reactivity of pulmonary arteries to inhaled NO, since the percent changes were similar among different degrees of PH and pulmonary vascular remodeling. The average percent reduction of mean PAP was 13% irrespective of inhaled NO concentration between 0.1 and 2.0 ppm. These animal studies formed the background for low-dose NO inhalation in postoperative cardiac patients [25], patients with pulmonary fibrosis [26, 30], patients with acute lung injury [29], and patients with chronic obstructive pulmonary disease [27]. Clinical experience showed that the maximum response is obtained with lower doses by improving oxygenation rather than treating pulmonary hypertension [5, 7]. The dose-response relationship might vary daily in any one patient and between patients [5, 7].

Effects of inhaled NO on Arterial Oxygenation (fig. 3)

The poorly ventilated alveolus has alveolar hypoxia constricting the pulmonary vasculature; this is termed hypoxic pulmonary vasoconstriction (HPV). One of the mechanisms of HPV is closing of the potassium channel by hypoxia. HPV reduces the blood flow in poorly ventilated lung units, which corrects the mismatch in the

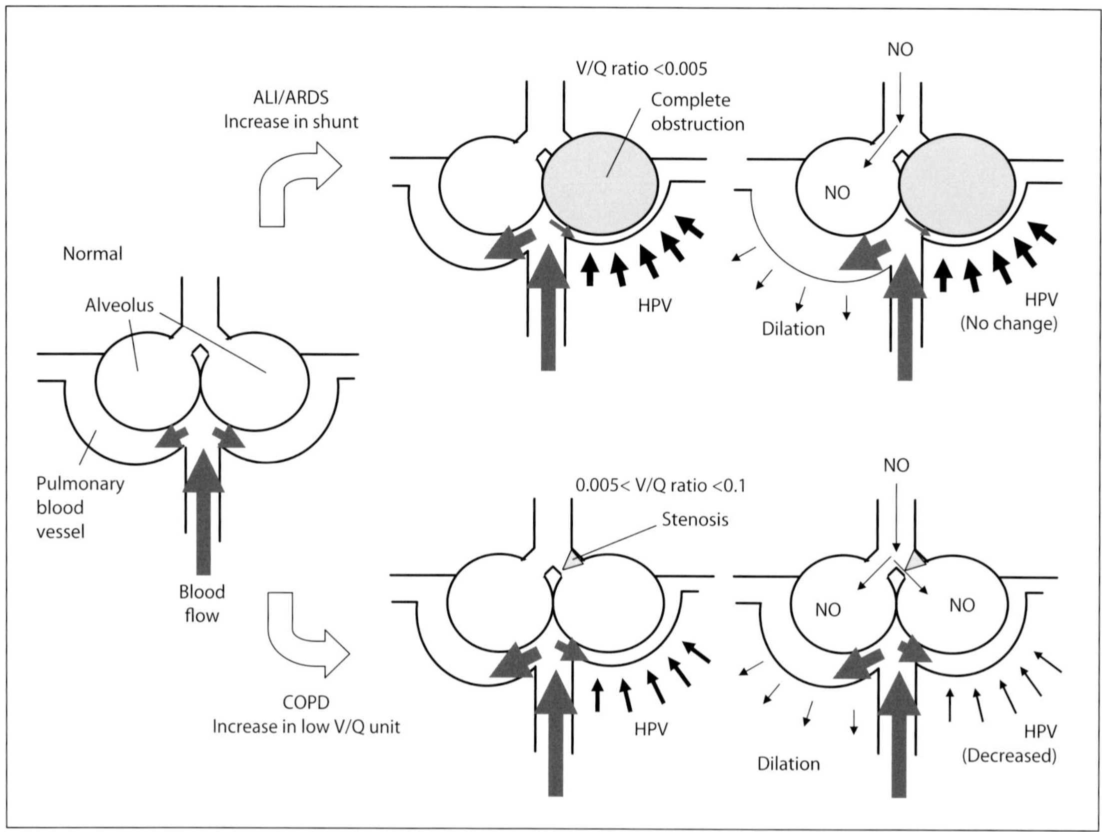

Fig. 3. Effect of NO on HPV in ARDS and COPD. In a lung unit with very low V/Q (<0.005) area, i.e. intrapulmonary shunt, inhaled NO cannot reach the constricted vasculature of a very low V/Q area because of no ventilation, so the inhaled NO does not reduce hypoxic pulmonary vasoconstriction (HPV) of the unit. In contrast, in a lung unit with low V/Q area (0.005 < V/Q < 0.1), NO enters these poorly ventilated units and partly reverses HPV. Very low V/Q area (intrapulmonary shunt) is high in ARDS, while low V/Q area is high in COPD.

ventilation-perfusion ratio. Since poorly ventilated units have low ventilation, the V/Q ratio becomes low. To correct this decreased V/Q ratio toward normal, perfusion (Q) should also decrease. To decrease the perfusion, vasoconstriction occurs spontaneously by HPV. Intrapulmonary shunt is defined as a very low V/Q ratio <0.005. Low V/Q area is defined as V/Q ratio 0.005–0.1, normal V/Q ratio is 0.1–10, and V/Q ratio >100 is dead space.

In the lung with a high intrapulmonary shunt, inhaled NO indirectly enhances the effect of HPV, i.e. reduction of blood flow in areas with a very low V/Q ratio, by dilating the vasculature in well-ventilated units. NO enters the ventilated unit and dilates the pulmonary vasculature where inhaled NO reaches, increasing the blood flow in the ventilated unit more than without NO inhalation. In patients with ALI/ARDS, intrapulmonary shunt (a very low V/Q area) is increased, while areas with low V/Q ratio are increased in patients with COPD.

Clinical Application of NO Inhalation

Persistent Pulmonary Hypertension of Neonates

In persistent pulmonary hypertension of neonates, increased PVR results in a right-to-left shunt across the patent ductus arteriosus and foramen ovale inducing severe hypoxemia which is often treated with extracorporeal membrane oxygenation (ECMO). Inhaled NO reduces the right-left shunt and improves arterial oxygenation [12, 13]. A subsequent randomized controlled trial showed that NO inhalation reduced the need for ECMO although it did not change the survival rate [34], leading to NO inhalation becoming a registered therapy for neonatal hypoxic respiratory failure associated with PH in the United States and Europe.

ALI/ARDS

In patients with acute lung injury (ALI)/acute respiratory distress syndrome (ARDS) inhaled NO reduced intrapulmonary shunt and improved arterial oxygenation [14]. A subsequent randomized controlled trial (RCT) [31] and meta-analysis of 12 RCTs of patients with ALI/ARDS showed that inhaled NO had no significant effect on hospital mortality, duration of mechanical ventilation, or ventilator-free days, while it improved arterial oxygenation for 1–4 days, but also increased the risk of developing renal dysfunction [32]. Routine use of inhaled NO in patients with ALI/ARDS is currently not recommended [32]. However, as a rescue therapy to improve refractory hypoxemia, inhaled NO is administered. For example, NO inhalation was used in 20 (32%) out of 68 patients with novel influenza A (H1N1) ARDS in a 2009 pandemic [33].

COPD and Pulmonary Fibrosis

As previously described, NO cannot enter a nonventilated area of very low V/Q units. In contrast, in the lung with a low V/Q ratio, NO can reach both well-ventilated and poorly ventilated units (fig. 3). NO reduces HPV in poorly ventilated units which then show lower V/Q ratios because the blood flow of poorly ventilated units increases. In patients with chronic obstructive pulmonary disease (COPD), inhaled NO worsened the V/Q mismatch and arterial oxygenation in patients breathing air [28]. In contrast, combined therapy with NO and oxygen increased arterial oxygenation more than inhalation of oxygen alone [30]. Similar findings were obtained in patients with pulmonary fibrosis [29]. Subsequent RCT showed that 3-month long-term NO inhalation with oxygen decreased PAP and PVR without decreasing arterial oxygenation, and increased cardiac output [23].

Congestive Heart Failure

A RCT showed that 40 ppm NO inhalation improved the arterial oxygenation in patients with congestive heart failure and decreased the PVR [38].

Left Ventricular Assist Device

Relaxation of the pulmonary vasculature reduces the PVR, which reduces the afterload of the right ventricle. Reduction of right ventriclular afterload might increase cardiac output (CO). A RCT showed that inhaled 20 ppm NO reduced the mean PAP and increased left ventricular assist device flow [35].

High-Altitude Pulmonary Edema

Inhaled 40 ppm NO improved the oxygenation in subjects with high-altitude pulmonary edema [37].

Heart Transplantaion

Heart transplantation in patients with PH is often complicated by post-bypass right heart failure. A RCT showed that NO inhalation accelerated the weaning from the cardiopulmonary bypass [36].

Others

NO inhalation is performed in patients with ischemia-reperfusion injury and in pulmonary vasoreactivity testing in the cardiac catheterization laboratory for patients with congenital heart disease [6].

Effects of NO on the Development of Pulmonary Vascular Changes

For all conditions causing PH, the vascular changes include new muscularization of normally nonmuscular peripheral pulmonary arteries, medial hypertrophy of muscular arteries and increase in extracellular matrix protein. NO prevents the smooth muscle cell proliferation [8]. Prolonged administration of L-arginine, but not D-arginine, ameliorated the development of PH and pulmonary hypertensive vascular changes in both chronic hypoxia-induced and monocrotaline-induced PH rats, suggesting that L-arginine modified the endogenous NO production [40]. In contrast, continuous NO inhalation prevented the development of chronic hypoxia-induced PH [41], but not in monocrotaline-induced PH [22]. Endogenous NO might prevent the development of PH by its pulmonary vasodilatory action and anti-proliferative effect on pulmonary vascular smooth muscle cells. In contrast, inhaled NO might prevent the development of PH mainly by the vasodilatory action, explaining why inhaled NO failed to prevent the development of PH in the monocrotaline model. In chronic hypoxia-induced PH, pulmonary vasoconstriction precedes the vascular structural changes, whereas vascular structural changes precede the rise in pulmonary artery pressure in monocrotaline-induced PH. The discrepancy of the drug effect between these two pH models has been studied in terms of all-trans-retinoic acid (ATRA) which might increase NO synthesis. ATRA prevented the development of PH in the monocrotaline model but not in the chronic hypoxia-induced PH model [19, 42].

NO increases cyclic-GMP in the target cells. Atrial natriuretic peptide (ANP) via natriuretic receptor A also increases c-GMP. Pulmonary artery dilatation responses by ANP are enhanced, possibly via a post-transcriptional modulation of the receptor in chronic hypoxia-induced PH in rats [43]. ANP gene transfection intratracheally with hemagglutinating virus Japan (a murine parainfluenza virus) ameliorated the development of medial hypertrophy of small muscular pulmonary arteries in chronic hypoxia-induced PH [44]. Thus, modulation to increase cGMP would be of benefit in preventing the development of vascular changes in PH.

Effect of Inhaled NO on Recovery of PH

As described above, inhaled NO prevents the development of PH and hypertensive pulmonary vascular changes. How about the effect of inhaled NO on the regression of pulmonary hypertensive structural changes? In chronic hypoxia-induced PH in rats, vascular changes regressed during recovery in air after chronic hypoxia, which was not facilitated by concomitant continuous NO inhalation in air [45]. The effect of prolonged NO inhalation on regression of pulmonary vascular remodeling is still unclear, but a first RCT in patients with COPD revealed a significant reduction in PVR after 3 months of NO pulsed inhalation therapy [23]. Continuous inhaled NO improves dyspnea in patients with idiopathic PH [24]. After surgical repair of congenital heart disease with PH, hypertensive vascular changes regress. If inhaled NO hastens this recovery process, postoperative NO inhalation would benefit patients by not only reducing PAP and improving arterial oxygenation but also by improving hypertensive vascular remodeling.

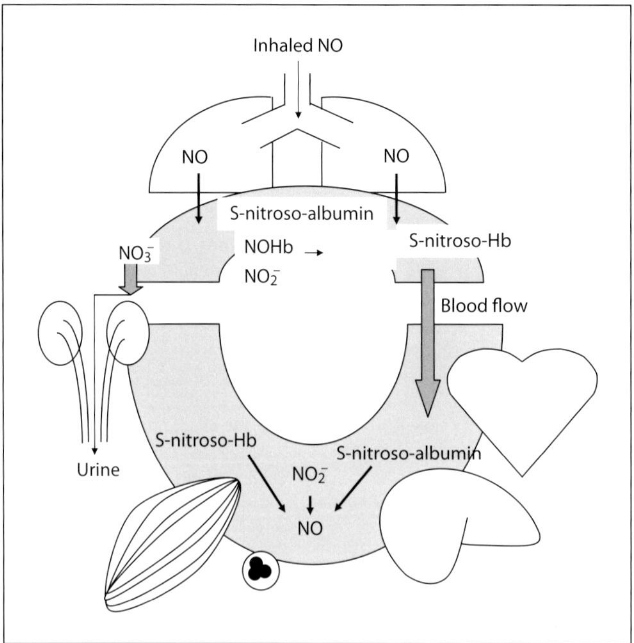

Fig. 4. Recycling of inhaled NO in remote organs other than lung. Inhaled NO diffuses in bloodsteam partly in the form of NO containing intermediates such as nitrite (NO_2^-), S-nitrosothiols , or C- and N-nitrosamines. These might circulate to the remote organs and act directly or after being recycled to NO. Nitrate (NO_3^-) is excreted in the urine.

Since endogenous NO production is impaired after prolonged NO administration [46], it was thought that prolonged NO inhalation might also impair the endogenous NO production in the pulmonary vasculature of PH. This hypothesis was not proved because continuous NO inhalation (10 and 40 ppm for 10 days) did not affect the impaired endogenous NO-cGMP relaxation cascade in isolated pulmonary arteries from rats during the recovery period after chronic hypoxia [18]. The mechanisms of a rebound increase in PAP after cessation of NO inhalation could not be explained from this result.

Effect of Inhaled NO on Remote Organ (fig. 4)

Inhaled NO 80 ppm during only the operative period of liver transplantation decreased the length of hospital stay and improved liver function, suggesting a reduction of ischemia/reperfusion injury [47]. Inhaled NO combines mainly with oxyhemoglobin to form methemoglobin and nitrate, and to a lesser extent with oxygen to form nitrite (NO_2^-), with thiol to form S-nitrosothiols such as S-nitrosoalbumin, and with amine to form C- or N-nitrosamines [7, 47]. Reduced-sulfur group (–SH) is termed thiol and is present in more than 100 proteins including hemoglobin and albumin [7]. These NO-containing intermediates might circulate to the remote organs and act directly or after being recycled to NO [47].

References

1 Furchigott RF, Zawadzki JV: The obligatory role of endothelial cells in the relaxation of arterial smooth muscle by acetylcholine. Nature 1980;288:373–376.

2 Ignarro LJ, Buga GM. Wood KS, Byrns RE, Chaudhuri G: Endothelium-derived relaxing factor produced and released from artery and vein is nitric oxide. Proc Natl Acad Sci USA 1987;84:9265–9269.

3 Palmer RM, Rerrige AG, Moncada S: Nitric oxide release accounts the biological activity of endothelium-derived relaxing factor. Nature 1987;327: 524–526.

4 Stamler JS, Singel DJ, Loscalzo J: Biochemistry of nitric oxide and its redox-activated forms. Science 1992;258:1898–1902.

5 Steudel W: Inhaled nitric oxide: basic biology and clinical applications. Anesthesiology 1999;91:1090–1121.

6 Ichinose F, Roberts JD, Zapol WM: Inhaled nitric oxide: a selective pulmonary vasodilator current uses and therapeutic potential. Circulation 2004;109: 3106–3111.

7 Griffiths MJ, Evans TW: Inhaled nitric oxide therapy in adults. N Engl J Med 2005;353:2683–2695.

8 Grag UC, Hassid A: Nitric oxide-generating vasodilators and 8-bromo-cyclic guanosine monophosphate inhibit mitogenesis and proliferation of cultured rat smooth muscle cells. J Clin Invest 1989; 83:1774–1777.

9 Yoshida K, Kasama K, Kitabatake M, Imai M: Biotransformation of nitric oxide, nitrite and nitrate. Int Arch Occup Environ Health 1983;52:103–115.

10 Frostell C, Fratacci MD, Wain JC, Jones R, Zapol WM: Inhaled nitric oxide, a selective pulmonary vasodilator reversing hypoxic pulmonary vasoconstriction. Circulation 1991;83:2038–2047.

11 Pepke-Zaba J, Higenbottam TW, Dinh-Xuan AT, Stone D, Wallwork J: Inhaled nitric oxide as a cause of selective pulmonary vasodilatation in pulmonary hypertension. Lancet 1991;338:1173–1174.

12 Roberts JD, Polaner DM, Lang P, Zapol WM: Inhaled nitric oxide in persistent pulmonary hypertension of the newborn. Lancet 1992;340:818–819.

13 Kinsella JP, Neish SR, Shaffer E, Abman SH: Low-dose inhalational nitric oxide in persistent pulmonary hypertension of the newborn. Lancet 1992;340: 819–820.

14 Rossaint R, Falke K, Lopez F, Slama K, Pison U, Zapol WM: Inhaled nitric oxide for the adult respiratory distress syndrome. N Engl J Med 1993;328: 399–405.

15 Dinh-Xuan AT, Higenbottam TW, Clelland CA, Pepke-Zaba J, Cremona G, Butt AY, Large SR, Wells FC, Wallwork J: Impairment of endothelium-dependent pulmonary-artery relaxation in chronic obstructive lung disease. N Engl J Med 1991;324: 1539–1547.

16 Maruyama J, Maruyama K: Impaired nitric oxide-dependent responses and their recovery in hypertensive pulmonary arteries of rats. Am J Physiol 1994;35:H2476-H2488.

17 Maruyama J, Yokochi A, Maruyama K, Nosaka S: Acetylcholine-induced endothelium-derived contracting factor in hypoxic pulmonary hypertensive rats. J Appl Physiol 1999;86:1687–1695.

18 Maruyama J, Jiang BH, Maruyama K, Takata M, Miyasaka K: Prolonged nitric oxide inhalation during recovery from chronic hypoxia does not decrease nitric oxide-dependent relaxation in pulmonary arteries. Chest 2004;125: 1919–1925.

19 Zhang E, Jiang B, Yokochi A, Maruyama J, Mitani Y, Ma N, Maruyama K: Effect of all-trans-retinoic acid on the development of chronic hypoxia-induced pulmonary hypertension. Circ J 2010;74:1696–1703.

20 Jiang BH, Maruyama J, Yokochi A, Mitani Y, Maruyama K: A novel inhibitor of inducible nitric oxide synthase, ONO-1714, does not ameliorate hypoxia-induced pulmonary hypertension in rats. Lung 2007;185:303–308.

21 Katayama Y, Hatanaka K, Hayashi T, Onoda K, Yada I, Namikawa S, Yuasa H, Kusagawa K, Maruyama K, Kitabatake M: Effects of inhaled nitric oxide in rats with chemically induced pulmonary hypertension. Respir Physiol 1994;94:301–307.

22 Maruyama J, Maruyama K, Mitani Y, Kitabatake M, Yamauchi T, Miyasaka K: Continuous low-dose NO inhalation does not prevent monocrotaline-induced pulmonary hypertension in rats. Am J Physiol 1997;272:H517–H524.

23 Vonbank K, Ziesche R, Higenbottam TW, Stiebellehner L, Petkov V, Schenk P, Germann P, Block LH: Controlled prospective randomized trial on the effects on pulmonary haemodynamics of the ambulatory long term use of nitric oxide and oxygen in patients with severe COPD. Thorax 2003;58: 289–293.
24 Perez-Penate G, Julia-Serda G, Pulido-Duque JM, Gorriz-Gomez E, Cabrera-Navarro P: One-year continuous inhaled nitric oxide for primary pulmonary hypertension. Chest 2001;119:970–973.
25 Simpo H, Mitani Y, Tanaka J, Mizumoto T, Onoda K, Tani K, Yuasa H, Yada I, Maruyama K: Inhaled low-dose nitric oxide for postoperative care in patients with congenital heart defects. Artif Organs 1997;21:10–13.
26 Maruyama K, Kobayashi H, Taguchi O, Chikusa H, Muneyuki M: Higher doses of inhaled nitric oxide might be less effective in improving oxygenation in a patient with interstitial pulmonary fibrosis. Anesth Analg 1995;81:210–211.
27 Maruyama K, Takeuchi M, Chikusa H, Muneyuki M: Reduction of intrapulmonary shunt by low-dose inhaled nitric oxide in a patient with late-stage respiratory distress associated with paraquat poisoning. Intensive Care Med 1995;21:778–779.
28 Barbera JA, Roger N, Roca J, Rovira I, Higenbottam TW, Rodriguez-Roisin R: Worsening of pulmonary gas exchange with nitric oxide inhalation in chronic obstructive pulmonary disease. Lancet 1996;347: 436–440.
29 Yoshida M, Taguchi O, Gabazza EC, Yasui H, Kobayashi T, Kobayashi H, Maruyama K, Adachi Y: The effect of low-dose inhalation of nitric oxide in patients with pulmonary fibrosis. Eur Respir J 1997; 10:2051–2054.
30 Yoshida M, Taguchi O, Gabazza EC, Kobayashi T, Yamakami T, Kobayashi H, Maruyama K, Shima T: Combined inhalation of nitric oxide and oxygen in chronic obstructive pulmonary disease. Am J Respir Crit Care Med 1997;155:526–529.
31 Dellinger RP, Zimmerman JL, Taylor RW, Straube RC, Hauser DL, Criner GJ, Davis K, Hyers TM, Papadakos P, Inhaled Nitric Oxide in ARDS Study Group: Effects of inhaled nitric oxide in patients with acute distress syndrome: results of randomized phase II trial. Crit Care Med 1998;26:15–23.
32 Adhirari NKJ, Burns KEA, Friedrich JO, Granton JT, Cook DJ, Meade MO: Effect of nitric oxide on oxygenation and mortality in acute lung injury: systematic review and meta-analysis. BMJ 2007;334: 779–782.
33 The Australia and New Zealand Extracorporeal Membrane oxygenation (ANZ ECMO) Influenza Investigators: Extracorporeal membrane oxygenation for 2009 influenza A(H1N1) acute respiratory distress syndrome. JAMA 2009;302:1888–1895.
34 Neonatal Inhaled Nitric Oxide Study Group: Inhaled nitric oxide in full-term and nearly-full term infants with hypoxic respiratory failure. N Engl J Med 1997;336:597–604.
35 Argenziano M, Choudhri AF, Moazami N, Rose EA, Smith CR, Levin HR, Smerling AJ, Oz MC: Randomized, double-blind trial of inhaled nitric oxide in LVAD recipients with pulmonary hypertension. Ann Thorac Surg 1998;65:340–345.
36 Rajek A, Pernerstorfer T, Kastner J, Mares P, Grabenwoger M, Sessler DI, Grubhofer G, Hiesmayr M: Inhaled nitric oxide reduced pulmonary vascular resistance more than prostaglandin E1 during heart transplantation. Anesth Analg 2000;90:523–530.
37 Scherrer U, Vollenweider L, Delabays A, Savcic M, Eichenberger U, Kleger GR, Fikrle A, Ballmer P, Nicod P, Bartsch P: Inhaled nitric oxide for high-altitude pulmonary edema. N Engl J Med 1996;334: 624–629.
38 Matsumoto A, Momomura S, Sugiura S, Fujita H, Aoyagi T, Sata M, Omata M, Hirata Y: Effect of inhaled nitric oxide on gas exchange in patients with congestive heart failure: a randomized, controlled trial. Ann Intern Med 1999;130:40–44.
39 Jiang BH, Maruyama J, Yokochi A, Amano H, Mitani Y, Maruyama K: Correlation of inhaled nitric-oxide induced reduction of pulmonary artery pressure and vascular changes. Eur Respir J 2002;20: 52–58.
40 Mitani Y, Maruyama K, Sakurai M: Prolonged administration of L-arginine ameliorates chronic pulmonary hypertension and pulmonary vascular remodeling in rats. Circulation 1997;96:689–697.
41 Kouyoumdjian C, Adnot S, Levame M, Eddahibi H, Bousbaa H, Raffestin B: Continuous inhalation of nitric oxide protects against development of pulmonary hypertension in chronically hypoxic rats. J Clin Invest 1994;94:578–584.
42 Preston IR, Tang G, Tilan JU, Hill NS, Susuki YJ: Retinoid and pulmonary hypertension. Circulation 2005;111:782–790.
43 Mitani Y, Maruyama J, Yokochi A, Maruyama K, Yoshimoto T, Naruse M, Sakurai M: Modulated vasodilatory responses to natriuretic peptides in rats exposed to chronic hypoxia. Eur Respir J 2000; 15:400–406.

44 Mitani Y, Maruayma J, Jiang BH, Sawada H, Simpo H, Imanaka-Yoshida K, Kaneda Y, Komada Y, Maruyama K: Atrial natriuretic peptide gene transfection with a novel envelope vector system ameliorates pulmonary hypertension in rats. J Thorac Cardiovasc Surg 2008;136:142–149.
45 Jiang BH, Maruyama J, Yokochi A, Iwasaki M, Amano H, Mitani Y, Maruyama K: Prolonged nitric oxide inhalation fails to regress hypoxic vascular remodeling in rat lung. Chest 2004;125:2247–2252.
46 Buga GM, Griscavage JM, Rogers NF, Ignarro LJ: Negative feedback regulation of endothelial cell function by nitric oxide. Circ Res 1993;73:808–812.
47 Lang JD, Teng X, Chumley P, Crawford JH, Isbell TS, Chacko BK, Liu Y, Jhala N, Crowe DR, Smith AB, Cross RC, Frenette L, Kelley EE, Wilhite DW, Hall CR, Page GP, Fallon MB, Bynon JS, Eckhoff DE, Patel RP: Inhaled NO accelerates restoration of liver function in adults following orthotopic liver transplantation. J Clin Invest 2007;117:2583–2591.
48 Maruyama K, et al: Nitric oxide in pulmonary circulation. Jpn J Intens Care Med 1994;18:1049–1057.

Kazuo Maruyama
Department of Anesthesiology and Critical Care Medicine,
Mie University School of Medicine 2–174 Edobashi,
Tsu Mie 514-8507 (Japan)
Tel. +81 59 231 5031, E-Mail k-maru@clin.medic.mie-u.ac.jp

Yoshikawa T, Naito Y (eds): Gas Biology Research in Clinical Practice.
Basel, Karger, 2011, pp 56–64

Inhaled Nitric Oxide and Clinical Application

Nobuaki Shime · Miho Inoue · Satoru Hashimoto

Department of Anesthesiology and Intensive Care, Kyoto Prefectural University, School of Medicine, Kyoto, Japan

Abstract

Inhaled nitric oxide is a potent pulmonary vasodilator, having clinically useful properties in improving pulmonary gas exchange and reducing pulmonary vascular resistance. It was officially approved for the clinical use in neonatal persistent pulmonary hypertension (PPHN) and has been utilized worldwide since year 1999. Since then, a significant reduction was observed in the rate of application of extracorporeal membrane oxygenation, a costly and invasive procedure, in the treatment of PPHN except for congenital diaphragmatic hernia. For the other various purposes, however, clinical evidence has not yet been confirmed. Clinical significance in prophylactic or therapeutic use for avoiding pulmonary hypertensive crisis (PHC) in children after congenital heart surgery has not yet been clarified, while the possibility remains of rescue use for a lethal PHC attack. Similarly, despite the transient improvement in arterial oxygenation in patients with acute lung injury/acute respiratory distress syndrome, it is not associated with an improvement in clinically significant outcome including mortality. Further studies are clearly needed to confirm its clinical usefulness in various clinical settings, taking into consideration the side effects and high costs as well as for direct comparison or combination with other recently developed potent pulmonary vasodilators including sildenafil or bosentan.

Nitric oxide (NO) is a colorless, odorless gas commonly known as an air pollutant. It is formed when nitrogen burns, as occurs in automobile engines and waste incineration plants. Endogenous NO has diverse physiological functions. In the cardiovascular system, vascular endothelial cells use NO to signal vascular smooth muscle cells to relax, thus resulting in vasodilatation and increased blood flow. Vasodilators such as nitroprusside and nitroglycerine release NO and mimic this effect. Among various physiological effects, vasodilating effect of NO in the lung has been studied extensively and applied in the treatment of pulmonary and cardiac conditions.

This chapter focuses on the mechanisms of the action of inhaled NO (iNO) and its clinical applications, with particular attention to persistent pulmonary hypertension of the newborn (PPHN), pulmonary hypertension after cardiac surgery and acute lung injury/acute respiratory distress syndrome (ALI/ARDS).

Mechanism of Action of Inhaled Nitric Oxide in the Lung

The vasodilating effect of NO in the lung is mediated by a cascade of intracellular signals. NO, either synthesized endogenously from L-arginine and oxygen by NO synthase in vascular endothelial cells or administered exogenously by inhalation, diffuses into the adjacent vascular smooth muscle cells and activates soluble guanylate cyclase. Activated guanylate cyclase catalyses the conversion of guanosine triphosphate (GTP) to cyclic guanosine 3′,5′-monophosphate (cGMP). cGMP activates protein kinase G, which ultimately leads to relaxation of the vascular smooth muscle. The action of cGMP is terminated by its hydrolysis to GMP by phosphodiesterases (PDEs). Among the isozymes of PDEs, PDE 5 plays a major role in vascular smooth muscle cells. Sildenafil, a drug currently indicated to alleviate pulmonary hypertension, inhibits PDE 5 and exerts its vasodilating effect through the same cGMP-mediated pathway. The action of NO administered by inhalation is short-lived and highly selective to lung because it is immediately scavenged by red blood cells once NO reaches the bloodstream. NO combines with oxyhemoglobin and produces methemoglobin and nitrate (NO_3^-). It also reacts with oxygen and water and produces nitrogen dioxide and nitrite (NO_2^-), respectively, which in turn interacts with oxyhemoglobin and forms methemoglobin and nitrate. More than 70% of inhaled NO is excreted in the urine as nitrate.

Inhaled NO improves oxygenation of arterial blood by dilating vessels in well-ventilated areas in the lung, diverting intrapulmonary blood flow from areas with low partial pressure of oxygen to areas with high partial pressure of oxygen and decreasing intrapulmonary shunting. This topical effect is a great advantage of iNO over oral and intravenous vasodilators which decrease oxygenation of arterial blood by causing diffuse pulmonary vasodilatation and increasing intrapulmonary shunting (i.e. ventilation perfusion mismatching). Oral and intravenous vasodilators also dilate systemic vessels and cause systemic hypotension. In addition to oxygenation of blood, iNO has favorable effects on cardiopulmonary hemodynamics. It reduces pulmonary vascular resistance and right ventricular afterload and ameliorates right ventricular failure.

Methods of Administration

NO gas is supplied in nitrogen or other inert gases in cylinders because NO readily reacts with oxygen to form NO_2, which irritates airway mucosa and causes pulmonary edema at high concentration. NO_2 is produced in proportion to the square of the concentrations of NO, the fraction of inspired oxygen (F_iO_2) and residence time of NO in oxygen [1]. To avoid the toxicity of NO_2, it is important to deliver NO in a way to minimize the exposure of NO to oxygen in the breathing circuit and accurately measure inhaled NO, NO_2 and O_2 by sampling the gas as close to the patient as possible.

In the early years of NO use, one of the technical challenges in administering NO was maintaining a constant concentration of NO in the breathing circuit where gas flow changes throughout the respiratory cycle. Commercially available NO delivery systems such as INOvent® and INOmax DS® NO delivery systems (Datex-Ohmeda, Madison, Wisc., USA) are designed to measure the mass of the gas flowing through the inspiratory limb of the ventilator circuit and deliver NO proportionally to the mass to provide a constant concentration of NO. The NO delivery systems also continuously monitor inhaled NO, NO_2 and O_2.

One of the complications of iNO therapy is methemoglobinemia. Methemoglobin is incapable of reversibly binding and transporting oxygen. High concentrations of methemoglobin can result in impaired oxygen delivery to peripheral tissues. Methemoglobin level in blood should to be monitored and maintained in the normal range. This complication is uncommon at a NO dose less than 20 ppm because methemoglobin reductase rapidly converts methemoglobin to hemoglobin.

Another complication is rebound pulmonary hypertension. A slow, stepwise reduction of iNO dose by 1 ppm is recommended after decreasing the dose to 5 ppm (table 1) [2].

Clinical Application

Inhaled NO was approved for the treatment of PPHN in term and near-term infants by the US Food and Drug Administration in 1999, by the European Medicine Evaluation Agency and European Commission in 2001, and by the Ministry of Health, Labor and Welfare, Japan, in 2009. In addition to PPHN, iNO has been used for the treatment of pulmonary hypertension after congenital heart surgery, right ventricular failure after orthotopic heart transplantation, primary pulmonary hypertension and ALI/ARDS in an effort to improve oxygenation or decreased pulmonary vascular resistance [3, 4]. This section highlights the application of iNO in the treatment of PPHN, pulmonary hypertension after cardiac surgery and ALI/ARDS.

Persistent Pulmonary Hypertension of Newborn

Persistent pulmonary hypertension of newborn (PPHN) occurs when normal transition from fetal to postnatal circulation fails to occur. In utero, the placenta is the organ for gas exchange. Because pulmonary vascular resistance is high, most blood from the right ventricle crosses the ductus arteriosus to the aorta and bypasses the lung. After birth, transitional changes in the lung such as mechanical distension with inhaled air, increase in oxygen tension and decrease in carbon dioxide tension trigger a rapid fall in pulmonary vascular resistance. Blood flow from the right ventricle to the lung increases dramatically and the lung starts serving as the organ for

Table 1. Summary of inhaled NO therapy

Indication	Treatment of term and near-term (>34 weeks) neonates with hypoxic respiratory failure associated with clinical or echocardiographic evidence of pulmonary hypertension
Dosage and administration	The recommended dose is 20 ppm In patients whose oxygenation improved with 20 ppm, decrease dose to 5 ppm within the first 4–24 h of treatment Slow stepwise reduction to 1 ppm before discontinuation There is no evidence of benefit with >20 ppm Consider discontinuing iNO when F_iO_2 <60% and PaO_2 >60 mm Hg Transient requirement for increased F_iO_2 is common after discontinuation of iNO Consider restarting iNO if prolonged need for increased F_iO_2 15 to 20% after discontinuation
Contraindication	Treatment of neonates known to be dependent on right-to-left shunting of blood
Warnings and precautions	Monitor methemoglobin, inspired NO_2 and PaO_2 for methemoglobinemia Elevated inspired NO_2 level Rebound pulmonary hypertension Abrupt discontinuation may cause increase in pulmonary artery pressure, decrease in PaO_2 and decrease in systemic blood pressure Sildenafil prevents rebound pulmonary hypertension after withdrawal of iNO Adverse effects due to high dose of iNO should be treated initially by decreasing iNO dose
Off-label use	Inhaled NO is commonly used for these conditions but the impact on clinical outcome is unclear for these conditions Premature newborn Congenital diaphragmatic hernia Acute respiratory failure in adults and children Congenital heart disease

gas exchange. In PPHN, pulmonary vascular resistance remains high after birth. A large amount of right ventricular output continues to bypass the lung, resulting in systemic arterial hypoxemia. The etiology of PPHN is heterogeneous and categorized into three groups: (1) the abnormally constricted pulmonary vasculature due to lung parenchymal diseases (e.g. meconium aspiration syndrome, respiratory distress syndrome, pneumonia); (2) the lung with normal parenchyma and remodeled pulmonary vasculature, also known as idiopathic persistent pulmonary hypertension of the newborn, or (3) the hypoplastic vasculature as seen in congenital diaphragmatic hernia (CDH) [5]. Response to iNO therapy can be different depending on the underlying conditions.

In term and near-term infants with PPHN, iNO is effective in improving oxygenation and decreasing the need for extracorporeal membrane oxygenation (ECMO) support [6–9]. According to a meta-analysis of randomized controlled trials (RCT), oxygenation improves in approximately 50% of infants with PPHN receiving NO. The oxygenation index decreases by a mean of 15.1 within 30–60 min after commencing therapy and PaO_2 increases by a mean of 53 mm Hg [10]. A practical guide in using iNO for neonatal PPHN is described in table 1.

Inhaled NO does not reduce the length of hospital stay and mortality, nor does it decrease the risk of developing chronic lung disease and neurodevelopmental impairment at 12 months of age [11, 12]. Initiation of iNO therapy in the early stage of PPHN at oxygenation index 15–25 does not improve these short- and long-term outcomes any further compared to initiation of the therapy at oxygenation index >25 [13].

Even without the long-term benefit, iNO is widely used in an attempt to avoid ECMO. The risk of major complications such as intracranial hemorrhage from systemic anticoagulation, thromboemobolism and infection and the cost of operating the ECMO machine are downsides of ECMO. A cost-effectiveness study of iNO in term and near-term newborns with hypoxemic respiratory failure showed that iNO is more effective in improving quality-adjusted survival and cheaper than ECMO [14]. Even in infants with severe respiratory failure who cannot avoid ECMO, iNO is effective in reducing the risk of cardiorespiratory arrest before undergoing ECMO [15].

Congenital Diaphragmatic Hernia

Infants with PPHN secondary to CDH have pulmonary hypoplasia. This potentially fatal abnormality seems to make this subgroup of PPHN patients less responsive to iNO therapy. In term and near-term infants with CDH and hypoxemic respiratory failure unresponsive to conventional therapy, iNO did not reduce the need for ECMO or mortality [6].

Prevention of Bronchopulmonary Dysplasia in Preterm Infants with Respiratory Distress Syndrome

In preterm infants with respiratory distress syndrome, studies on the effect of iNO in preventing bronchopulmonary dysplasia (BPD) have mixed results. BPD is a chronic lung disease commonly seen in preterm infants who received prolonged mechanical ventilation and oxygen to treat respiratory distress syndrome. It is defined as dependence to mechanical ventilation, continuous positive airway pressure or supplemental oxygen concentration of 30% and oxygen saturations of 90–96% at 35–37 weeks of postmenstrual age [16]. In infants with birth weight more than 1,000 g, the use of iNO for respiratory distress syndrome may decrease the incidence of BPD and death

[17, 18]. In very preterm infants born between 24 and 28 weeks plus 6 days of postmenstrual age, iNO does not improve survival without development of BPD or brain injury at 36 weeks of postmenstrual age [19]. Further studies are necessary to define subgroups of preterm infants who are likely to benefit from iNO therapy.

Pulmonary Hypertension after Cardiac Surgery

Pulmonary hypertension crisis (PHC), represented by acute hypoxemia combined with right ventricular failure due to sudden increases in pulmonary arterial resistance, is one of the most lethal complications occurring after surgical repair of congenital cardiac defect with left-to-right intra- and extracardiac shunt [20]. Traditional ways to alleviate the increases in pulmonary vascular resistance include reduction of stress response by administering narcotics, sedatives, and, occasionally, muscle relaxants, as well as the use of intravenous pulmonary vasodilators (nitroglycerin, nitroprusside, phosphodiesterase-III inhibitors, etc.) or the avoidance of hypoxemic pulmonary vasoconstriction by oxygenation and alkalosis. Those therapies, however, possesses significant hemodynamic side effects (hypoxemia and systemic hypotension due to vasodilators, or increased systemic vascular resistance due to hyperventilation-induced alkalosis, etc.), which is specifically problematic in hemodynamically unstable patients after cardiac surgery [21].

Inhaled NO, having the potential to selectively dilate pulmonary vessels, could be served as a possible therapeutic option for the increased pulmonary vascular resistance and PHC. Preliminary results have shown that iNO improves oxygenation and decreases pulmonary arterial pressure in patients with pulmonary hypertension after congenital heart surgery. In 2000, Miller et al. [22] conducted a relatively large-scale RCT to determine the clinical effects of iNO (n = 124) in infants with preoperative high pulmonary flow or pressure, and were undergoing corrective surgery for congenital heart disease. In the study, it was revealed that significant reduction of the occurrence of postoperative PHC and the time to fulfill extubation criteria was induced by prophylactic iNO. Another smaller RCT conducted by Day et al. [23] in 2000, however, did not show any benefit of iNO in reducing the rate of severe PHC, defined by dramatic increases in pulmonary arterial pressure exceeding systemic pressure, systemic hypotension, and hypoxemia. Differences in the eligibility criteria for the study, outcome assessment might affect the discrepancy between the results in the two RCTs. Given that no significant differences in clinically significant outcomes (duration of mechanical ventilation, duration of ICU stay, or mortality) in both of the studies, the prophylactic use of iNO for postoperative pulmonary hypertension should be cautiously applied individually, by concerning the pre- and postoperative pulmonary and systemic arterial pressures and ventricular function.

One small case-series have suggested that the use of iNO as the therapeutic, rescue use for lethal PHC, indicating the possibility of reducing the induction of ECMO

[24]. More confirmative studies, however, are needed to assess the effect of iNO as a therapeutic purpose. Current studies have suggested the combination use of iNO with other potent pulmonary vasodilators (sildenafil or bosentan) [25, 26]. Studies are needed to clarify whether iNO still have a role in the era of the evolving those newer clinically applicable pulmonary vasodilators.

Right Heart Bypass Surgery

Congenital heart surgery patients after right heart bypass, including Glenn shunt or Fontan-type surgery, are another possible population benefiting from the reduction in pulmonary vascular resistance with iNO. As the pulmonary blood flow is not delivered by the contraction of the right heart, and is passively regulated by the pulmonary vascular resistance, keeping lower pulmonary vascular resistance is the key to keep pulmonary blood flow and systemic cardiac output constant. Patients with higher pulmonary vascular resistance represented by an increased central venous pressure ≥15–20 mm Hg or transpulmonary pressure gradient ≥8–10 mm Hg could be the target for the use of iNO [27, 28]. No RCTs, however, to date have yet confirmed the clinical significance of the iNO in such populations.

Acute Lung Injury/Acute Respiratory Distress Syndrome

ALI/ARDS is the acute onset of hypoxemia in the absence of clinical signs of left arterial hypertension, characterized by bilateral infiltrates on chest radiographs. Many studies have evaluated the efficacy of iNO for the treatment of ALI/ARDS. According to a systematic review including 1,303 patients with ALI/ARDS from 14 RCTs, iNO transiently improves oxygenation in the first 24 h but does not reduce mortality [29]. Considering that progression of ARDS to multiorgan failure is the major cause of death, the improvement of oxygenation may not translate into decreased mortality. In addition, iNO may increase the risk of renal impairment. Therefore, iNO cannot be recommended for patients with ALI/ARDS.

Future Direction

Despite the potential of iNO in improving pulmonary gas exchange and reducing pulmonary vascular resistance, its clinical applicability has not yet been thoroughly confirmed by clinical trials with a higher level of quality and evidence. Moreover, the clinical utility is officially approved only for PPHN. Further studies, including large-scale RCTs, are clearly needed to determine its clinical usefulness by considering its side effects and high costs. Direct comparison, or combinations with newly developed, potent pulmonary vasodilators including sildenafil or bosentan, should be explored. Children with congenital cardiac defects, a major population among the off-label use, are the most possible candidates for the future trials.

References

1 Nishimura M, Hess D, Kacmarek RM, Ritz R, Hurford WE: Nitrogen dioxide production during mechanical ventilation with nitric oxide in adults: effects of ventilator internal volume, air versus nitrogen dilution, minute ventilation, and inspired oxygen fraction. Anesthesiology 1995;82:1246–1254.

2 Kinsella JP, Abman SH: Clinical approach to inhaled nitric oxide therapy in the newborn with hypoxemia. J Pediatr 2000;136:717–726.

3 George I, Xydas S, Topkara VK, Ferdinando C, Barnwell EC, Gableman L, Sladen RN, Naka Y, Oz MC: Clinical indication for use and outcomes after inhaled nitric oxide therapy. Ann Thorac Surg 2006; 82:2161–2169.

4 Taylor RW, Zimmerman JL, Dellinger RP, Straube RC, Criner GJ, Davis K Jr, Kelly KM, Smith TC, Small RJ: Low-dose inhaled nitric oxide in patients with acute lung injury: a randomized controlled trial. JAMA 2004;291:1603–1609.

5 Farrow KN, Fliman P, Steinhorn RH: The diseases treated with ECMO: focus on PPHN. Semin Perinatol 2005;29:8–14.

6 Anonymous: Inhaled nitric oxide and hypoxic respiratory failure in infants with congenital diaphragmatic hernia: the neonatal inhaled nitric oxide study group (NINOS). Pediatrics 1997;99:838–845.

7 Roberts JD Jr, Fineman JR, Morin FC 3rd, Shaul PW, Rimar S, Schreiber MD, Polin RA, Zwass MS, Zayek MM, Gross I, Heymann MA, Zapol WM: Inhaled nitric oxide and persistent pulmonary hypertension of the newborn: the inhaled nitric oxide study group. N Engl J Med 1997;336:605–610.

8 Davidson D, Barefield ES, Kattwinkel J, Dudell G, Damask M, Straube R, Rhines J, Chang CT: Inhaled nitric oxide for the early treatment of persistent pulmonary hypertension of the term newborn: a randomized, double-masked, placebo-controlled, dose-response, multicenter study. The I-NO/PPHN study group. Pediatrics 1998;101:325–334.

9 Clark RH, Kueser TJ, Walker MW, Southgate WM, Huckaby JL, Perez JA, Roy BJ, Keszler M, Kinsella JP: Low-dose nitric oxide therapy for persistent pulmonary hypertension of the newborn: clinical inhaled nitric oxide research group. N Engl J Med 2000;342:469–474.

10 Finer NN, Barrington KJ: Nitric oxide for respiratory failure in infants born at or near term. Cochrane Database Syst Rev 2006:CD000399.

11 Clark RH, Huckaby JL, Kueser TJ, Walker MW, Southgate WM, Perez JA, Roy BJ, Keszler M: Low-dose nitric oxide therapy for persistent pulmonary hypertension: 1-year follow-up. J Perinatol 2003;23: 300–303.

12 Lipkin PH, Davidson D, Spivak L, Straube R, Rhines J, Chang CT: Neurodevelopmental and medical outcomes of persistent pulmonary hypertension in term newborns treated with nitric oxide. J Pediatr 2002;140:306–310.

13 Konduri GG, Solimano A, Sokol GM, Singer J, Ehrenkranz RA, Singhal N, Wright LL, Van Meurs K, Stork E, Kirpalani H, Peliowski A: A randomized trial of early versus standard inhaled nitric oxide therapy in term and near-term newborn infants with hypoxic respiratory failure. Pediatrics 2004;113: 559–564.

14 Angus DC, Clermont G, Watson RS, Linde-Zwirble WT, Clark RH, Roberts MS: Cost-effectiveness of inhaled nitric oxide in the treatment of neonatal respiratory failure in the United States. Pediatrics 2003;112:1351–1360.

15 Fliman PJ, deRegnier RA, Kinsella JP, Reynolds M, Rankin LL, Steinhorn RH: Neonatal extracorporeal life support: impact of new therapies on survival. J Pediatr 2006;148:595–599.

16 Walsh MC, Yao Q, Gettner P, Hale E, Collins M, Hensman A, Everette R, Peters N, Miller N, Muran G, Auten K, Newman N, Rowan G, Grisby C, Arnell K, Miller L, Ball B, McDavid G: Impact of a physiologic definition on bronchopulmonary dysplasia rates. Pediatrics 2004;114:1305–1311.

17 Van Meurs KP, Wright LL, Ehrenkranz RA, Lemons JA, Ball MB, Poole WK, Perritt R, Higgins RD, Oh W, Hudak ML, Laptook AR, Shankaran S, Finer NN, Carlo WA, Kennedy KA, Fridriksson JH, Steinhorn RH, Sokol GM, Konduri GG, Aschner JL, Stoll BJ, D'Angio CT, Stevenson DK: Inhaled nitric oxide for premature infants with severe respiratory failure. N Engl J Med 2005;353:13–22.

18 Kinsella JP, Greenough A, Abman SH: Bronchopulmonary dysplasia. Lancet 2006;367:1421–1431.

19 Mercier JC, Hummler H, Durrmeyer X, Sanchez-Luna M, Carnielli V, Field D, Greenough A, Van Overmeire B, Jonsson B, Hallman M, Baldassarre J: Inhaled nitric oxide for prevention of bronchopulmonary dysplasia in premature babies (EUNO): a randomised controlled trial. Lancet 2010;376:346–354.
20 Hopkins RA, Bull C, Haworth SG, de Leval MR, Stark J: Pulmonary hypertensive crises following surgery for congenital heart defects in young children. Eur J Cardiothorac Surg 1991;5:628–634.
21 Morris K, Beghetti M, Petros A, Adatia I, Bohn D: Comparison of hyperventilation and inhaled nitric oxide for pulmonary hypertension after repair of congenital heart disease. Crit Care Med 2000;28: 2974–2978.
22 Miller OI, Tang SF, Keech A, Pigott NB, Beller E, Celermajer DS: Inhaled nitric oxide and prevention of pulmonary hypertension after congenital heart surgery: a randomised double-blind study. Lancet 2000;356:1464–1469.
23 Day RW, Hawkins JA, McGough EC, Crezee KL, Orsmond GS: Randomized controlled study of inhaled nitric oxide after operation for congenital heart disease. Ann Thorac Surg 2000;69:1907– 1912; discussion 1913.
24 Goldman AP, Delius RE, Deanfield JE, de Leval MR, Sigston PE, Macrae DJ: Nitric oxide might reduce the need for extracorporeal support in children with critical postoperative pulmonary hypertension. Ann Thorac Surg 1996;62:750–755.
25 Lepore JJ, Maroo A, Bigatello LM, Dec GW, Zapol WM, Bloch KD, Semigran MJ: Hemodynamic effects of sildenafil in patients with congestive heart failure and pulmonary hypertension: combined administration with inhaled nitric oxide. Chest 2005;127:1647–1653.
26 Steinhorn RH, Kinsella JP, Pierce C, Butrous G, Dilleen M, Oakes M, Wessel DL: Intravenous sildenafil in the treatment of neonates with persistent pulmonary hypertension. J Pediatr 2009;155: 841–847, e841.
27 Gamillscheg A, Zobel G, Urlesberger B, Berger J, Dacar D, Stein JI, Rigler B, Metzler H, Beitzke A: Inhaled nitric oxide in patients with critical pulmonary perfusion after Fontan-type procedures and bidirectional Glenn anastomosis. J Thorac Cardiovasc Surg 1997;113:435–442.
28 Yoshimura N, Yamaguchi M, Oka S, Yoshida M, Murakami H, Kagawa T, Suzuki T: Inhaled nitric oxide therapy after Fontan-type operations. Surgery today 2005;35:31–35.
29 Afshari A, Brok J, Moller AM, Wetterslev J: Inhaled nitric oxide for acute respiratory distress syndrome (ARDS) and acute lung injury in children and adults. Cochrane Database Syst Rev 2010;7:CD002787.

Nobuaki Shime, MD
Department of Anesthesiology and Intensive Care
Kyoto Prefectural University of Medicine
465 Kajii-cho, Kamigyo-ku
Kyoto 602-8566 (Japan)
Tel. +81 75 251 5633, Fax +81 75 251 5843, E-Mail shime@koto.kpu-m.ac.jp

Yoshikawa T, Naito Y (eds): Gas Biology Research in Clinical Practice.
Basel, Karger, 2011, pp 65–72

Hydrogen Sulfide in the Gastrointestinal Tract: Friend or Foe?

Chan Young Ock[a] · So Jung Han[b] · Ki-Seok Choi[a] · Eun-Hee Kim[a] · Ju Hyun Kim[b] · Ki-Baik Hahm[a,b]

[a]Laboratory of Translational Medicine, Gachon University, Lee Gil Ya Cancer and Diabetes Institute, and [b]Department of Gastroenterology, Gachon University Graduate School of Medicine, Incheon, Korea

Abstract

Hydrogen sulfide (H_2S) has been engaged in the reversible state of hypothermia, suspended animation-like states, and halitosis in vertebrates. In the gastrointestinal (GI) tract, H_2S has been put on the chopping board since a rather conflicting fact exists that inhibition of H_2S production causing obstacles in recovery from various animal models of inflammation, reperfusion injury, and compromised circulation or the possibility of therapeutic exploitation of H_2S or its donor compounds based on their anti-inflammatory and circulatory preservation. As a third gasotransmitter, together with nitric oxide and carbon monoxide, there are complex interactions between these mediators in their contribution to regulating cell function, vascular responses, and inflammatory reactions. In spite of these double-edged roles, many concerns have focused on H_2S as a potential therapeutic or a newer pathophysiological player. In this chapter, we describe the benevolence or notoriety of H_2S in the GI tract.

Small-molecular-weight gases including nitric oxide (NO), carbon monoxide (CO), and hydrogen sulfide (H_2S) constitute a unique class of biomaterials that are indispensable for maintaining the homeostasis of biological systems. They serve as a substance that readily conveys the signal from one site to another in an autocrine, paracrine or juxtacrine fashion since the gases are highly membrane-permeable. These gases exert their biological actions through interactions with proteins in multiple ways distinct from other signaling molecules. Interestingly, not only can different gases that share a similar chemical structure exert comparable biological actions but they often compete with and are antagonists to each other, showing that a single, small molecule such as a gas could affect various signaling pathways when exposed to an individual cell. Before there was enough understanding of the biologic roles of gases, we just thought of these gases as being either simple byproducts or waste material that needed to be discarded. However, with increasing evidence showing that they have more biologic roles than we realized – gases can easily penetrate the cell barrier and move faster than other signal molecules –, their impact on the whole body system seems to be far greater than that of conventional signaling molecules such as hormones and neurotransmitters [1, 2].

The first biogas that was seen to have a potent impact on the human body was NO. It was found by Drs. Furchgott, Murad, and Ignarro who were honored for their discovery by being awarded the Nobel Prize for medicine in 1998. Even though the studies of endothelium-derived relaxing factor (EDRF) and nitrovasodilators were discovered separately, two tales finally merged by finding the exact biological similarity of EDRF and NO. NO is synthesized from oxygen and L-arginine by NO synthase, and has major physiologic functions as a vasodilator, platelet aggregator, and neurotransmitter. It is known that NO interacts with various biomolecules that contain heme group-like guanylyl cyclase or cytochrome P450, and modulates the thiol status by interrupting the cysteine residues of proteins. As the second endogenous gas of interest, CO was discovered as an endogenously produced mediator and neurotransmitter. CO is synthesized during the metabolism of heme to biliverdin by heme oxidase-1 (HO-1). Generally known as a toxic gas produced by burning coal, CO asphyxia could cause serious damage since the affinity to hemoglobin is much higher with CO than with oxygen. However, increasing evidence has shown that CO can scavenge reactive oxygen species, regulate vascular muscle tone, or transmit neural signals to the brain, meaning that CO is a crucial mediator for maintaining physiological homeostasis [3, 4]. H_2S, the third gas of interest and importance, was also first known as a 'toxic gas' because for decades it was only known as a toxic environmental pollutant and principal offender of halitosis based on its strong odor of rotten eggs [5]. Even though, at the end of 1980s, endogenous hydrogen sulfide was found in the brain, it was suggested to be an artifact until Drs. Abe and Kimura described the enzymatic mechanism of H_2S production in the brain, its biological effects at physiological concentrations, and its specific cellular targets. H_2S is synthesized endogenously in various mammalian tissues by two pyridoxal-5′-phosphate-dependent enzymes responsible for metabolizing L-cysteine: cystathionine β-synthase (CBS) and cystathionine γ-lyase (CSE) [6]. The substrate of CBS and CSE, L-cysteine, is a sulfur-containing amino acid and can be derived from gastrointestinal (GI) sources or liberated from endogenous proteins. However, beyond this notoriety, increasing evidence has shown that H_2S could exert its beneficial physiological roles including the regulation of vasodilatation, attenuation of inflammation and modulation of gut signaling molecules, after which it became a key new therapeutic target, for instance, in the development of GI-safe NSAIDs, endotoxemia-induced gut injury, and GI motility disorders in the GI tract. In this chapter, we will focus on the role of H_2S in the GI system, providing both of its aspects: integrating the initiation and progression of GI diseases in a friendly or hostile manner.

Benevolent Role of H_2S Gas in the Gastrointestinal Tract

Administration of H_2S produced a 'suspended animation-like' metabolic status with hypothermia and reduced oxygen demand in pigs and mice, thus protecting from lethal hypoxia. This hypometabolic state, which resembles hibernation, induced by H_2S may contribute to tolerance against oxidative stress [7]. In addition, the antioxidant effects of H_2S can be explained by its effects on cytochrome *c* oxidase and mitochondrial functions, while its effects on gene expression may be related to actions on the the NF-κB and extracellular signal-regulated kinase pathways [8]. These antioxidant effects of H_2S are reported to be dependent on its concentration and the cellular status of the host. At micromolar concentrations, the cytoprotective effects of H_2S are usually produced by Na_2S or NaHS which can scavenge groups of reactive species including oxyradicals, peroxynitrite, hypochlorous acid and homocysteine [9]. Not does only H_2S exert anti-oxidant functions inside the cells, but it can also harmonize physiological interactions with other types of adjacent cells. These variable interactions are hard to be explained by one critical mechanism, but the possible major target of pharmacological effects of H_2S is the opening of potassium-opened ATP channels (K_{ATP} channels) [10]. Activating K_{ATP} channels results in vasodilatory, cardiac-protective and anti-nociceptive effects. Similar to these cardioprotective effects, since ischemic/reperfusion injury following vasoconstriction of vessels supplying the stomach mucosa is the main mechanism of gastric inflammation/ulceration caused by NSAIDs or *Helicobacter pylori*, vasodilation due to H_2S makes blood flow more sufficient thereby enhancing the gastric defense system. Interestingly, chemical ablation of sensory afferent neurons by capsaicin reversed the gastroprotective effects of NaHS, which means that the K_{ATP} channels in sensory neurons could be an important target of the gastroprotective effects of H_2S [11].

Based on these cytoprotective and beneficial biological actions of H_2S gas, Wallace et al. [12] have been studying the pharmacological possibilities of H_2S-releasing NSAIDs, which lowered stomach and small intestine toxicity as well as hematological toxicity compared with conventional NSAIDs, diclofenac. Since the induction of NSAIDs decreased CSE expression and H_2S synthesis, decreased H_2S level around the vessels supplying the gastric mucosa may enhance mucosal damage. An additional interesting finding is that inhibition of K_{ATP} channels by H_2S released from a H_2S donor reduced the leukocyte adherence to mesenteric venules induced by NSAIDs or aspirin. Moreover, the H_2S donor prevented many of the other pro-inflammatory effects of NSAIDs, including elevated levels of ICAM-1, LFA-1, and mucosal TNF-α expression. Since H_2S-releasing NSAIDs have been shown to reduce edema formation and leukocyte adherence to the vascular endothelium, inhibit pro-inflammatory cytokine synthesis, and increase the resistance of the gastric mucosa to injury and accelerate repair, anti-inflammatory drugs that are modified to release H_2S will exhibit improved efficacy and reduced toxicity. Such compounds have now been synthesized and shown to be markedly improved in many respects compared to the parent anti-inflammatory drugs, shedding new hope of solving the century-old problem of NSAID-associated GI injury [13, 14].

In an additional report dealing with the charitable role of H_2S besides its well-acknowledged H_2S-releasing role with NSAIDs, Wallace et al. [15] showed increased H_2S synthesis as well as enhanced expression of CSE and CBS in an acetic acid-induced acute ulcer model. Injection of H_2S donors after ulcer induction promoted ulcer healing in a dose-dependent manner. Distrutti et al. [16] investigated the role of H_2S in modulating nociception to colorectal distension (CRD), a model that mimics some features of irritable bowel syndrome, and found that H_2S inhibits nociception induced by CRD in both healthy and postcolitic rats, giving new anticipation that H_2S-releasing drugs might be beneficial in treating painful intestinal disorders. Moreover, H_2S can ameliorate H_2O_2-mediated oxidative stress in cultured normal rat gastric epithelial cells and in animal stress models. Besides of these benefits in gut, Jha et al. [17] published evidence that H_2S attenuated hepatic ischemia-reperfusion injury through upregulation of intracellular antioxidant and anti-apoptotic signaling pathways. Another study indicated that leukocytes might exert their anti-bacterial function very well in the presence of a high concentration of H_2S, though not with the low concentration. This phenomenon could explain the anti-bacterial activities of H_2S in the gingival pockets of periodontal diseases. Though tried in the cardiovascular system, adenovirus-mediated transfer of the H_2S-generating gene (CSE gene) may provide a novel therapeutic approach in treating vascular diseases linked to abnormal vascular remodeling like abdominal angina [18].

Notoriety Role of H_2S Gas in the Gastrointestinal Tract

Contrary to the above-mentioned beneficial actions of H_2S, the harmful effects of H_2S have also been reported. Multiple lines of evidence reveal that H_2S stimulates inflammation, i.e. H_2S may upregulate the inflammatory response via stimulation of immune cells. Reactive oxygen species (ROS) from activated neutrophils converted H_2S to sulfite, which may upregulate leukocyte adhesion and neutrophil functions. Li et al. [19] showed increased H_2S synthesis in LPS-induced inflammation in mice and further documented that inhibition of H_2S formation by DL-propargylglycine (PAG, CSE inhibitor) dramatically reduced the severity of LPS-induced inflammation in various organs. Zhang et al. [20] published that endogenous H_2S regulates leukocyte trafficking in cecal ligation and puncture-induced sepsis. Similarly, the correlations of H_2S with inflammatory diseases are also well defined in a model of pancreatitis. Bhatia et al. [21] clearly showed the role of H_2S in acute pancreatitis induced by caerulein, which could damage pancreas as well as lung. Increased levels of plasma H_2S, amylase, and pancreatic MPO were observed in the caerulein induction group, but PAG pretreatment decreased the above effects. Moreover, PAG pretreatment also ameliorated pancreatitis-associated lung injury assayed by lung

MPO activity. H_2S may mediate diseases associated with endotoxemia such as hemorrhagic shock.

Besides the implicating role of H_2S in leukocyte-mediated inflammation, it can induce endoplasmic reticulum stress, apoptosis [22], and sepsis and interplay with NO and thereby enhance inflammation and organ dysfunction [23]. The results drawn from our study throw light on a novel implication of H_2S in *H. pylori*-associated gastric inflammation and erosive changes (fig. 1). The plausible explanation why increased generation of H_2S accompanied with halitosis was noted in patients with *H. pylori* infection evidenced as the increased expression of the CBS and CSE genes, increased amounts of H_2S in gastric juices of patients with *H. pylori*-associated erosive change, correlation between H_2S and gastric inflammation, and attenuation of gastric inflammation through the lowering of H_2S levels. Increased leukocyte migration and cytokine expression as well as the expression of CBS and CSE was observed in Raw264.7 treated with the H_2S generator NaHS. The mitogen-activated protein kinase (MAPK) pathway is also activated and pretreatment of the MAPK inhibitor decreased the expression of CBS and CSE. These data lead to the conclusion that H_2S might be a villain in the GI tract and therefore its level should be lowered [24].

Conclusion: Great Pharmacological Potential of H_2S in the Gastrointestinal Tract

Each of the biologically active gases NO and CO is synthesized by enzymes which have been characterized biochemically and pharmacologically, and each acts, via well-established molecular targets, to effect physiological and/or pathophysiological functions within the body. Shared biological roles of the gases include regulation of vascular homoeostasis, modulation of gut inflammation, and chance of regeneration [25]. With the recent acknowledgement of H_2S as the third gaseous neurotransmitter, the term 'gasotransmitter' was introduced to characterize gases which act as neurally released transmitters. In summary, NO, CO and H_2S share distinct properties, which qualify them as gasotransmitters based on the facts that they are small molecules of gas, they are freely permeable across membranes and do not act via specific membrane receptors, they are synthesized endogenously and enzymatically on demand and their generation is regulated, they have well-defined specific functions at physiologically relevant concentrations, and their cellular effects may or may not be mediated by second messengers, but that these gasotransmitters have specific cellular and molecular targets. Due to their gaseous nature, NO, CO and H_2S are not stored within the cell in the classic presynaptic vesicles before they are released, but rather they are synthesized and released on demand, which distinguishes them from the classic neurotransmitters such as acetylcholine, norepinephrine, and even the peptide neurotransmitters. Medical gases are pharmaceutical gaseous molecules including traditional gases, such as oxygen and NO, as well as gases with recently discovered roles as biological messenger molecules, such as NO, CO and H_2S. Though gas therapy is

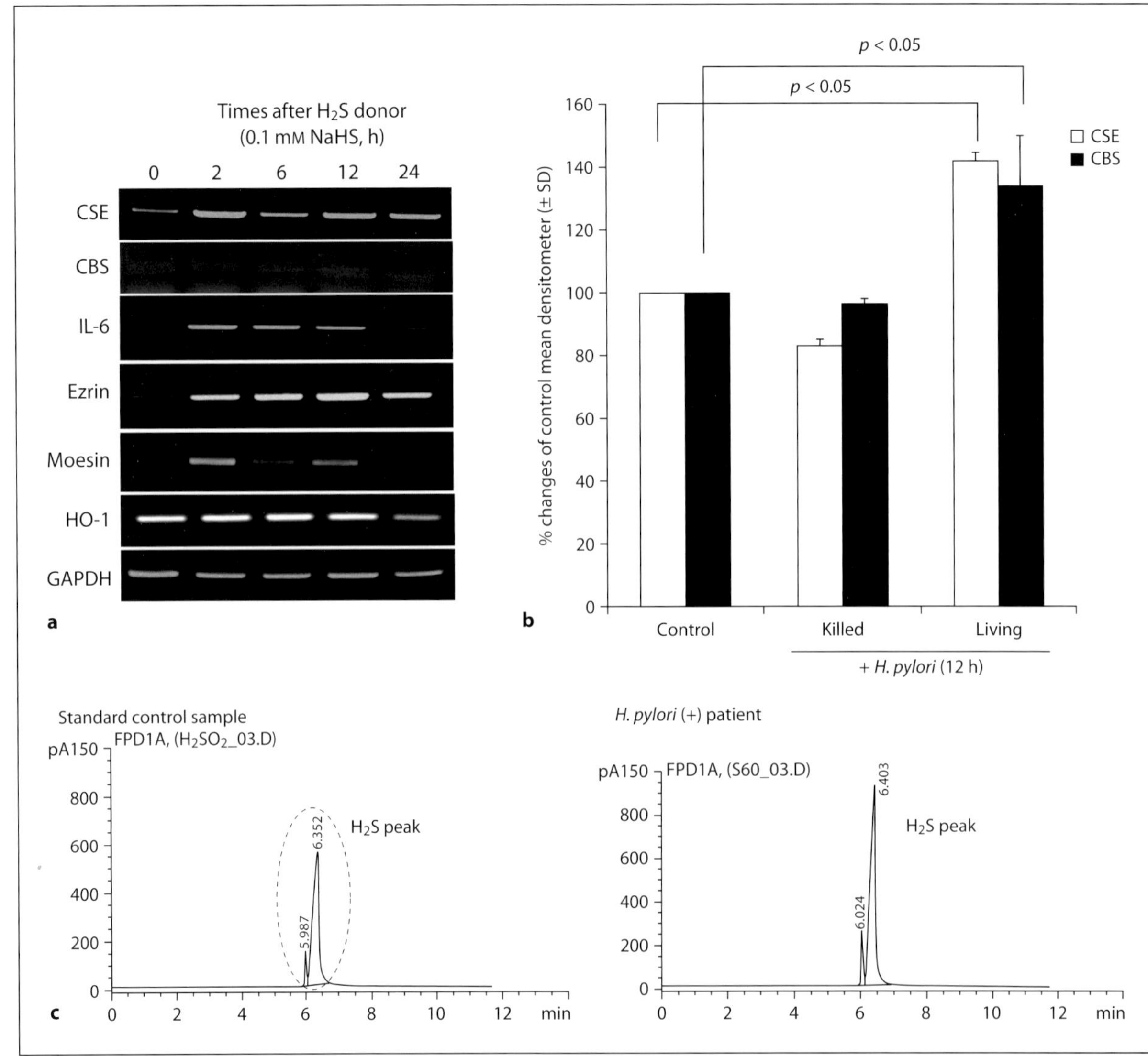

Fig. 1. H_2S generation with *H. pylori* infection [24]. **a** Changes in the expression of CSE, CBS, IL-6, ezrin, moesin and HO-1 at different time points after H_2S donor administration. H_2S generation were associated with a significant induction of the H_2S-generating enzyme, cytokine and leukocyte activation genes. **b** Change of CSE and CBS after *H. pylori* infection. Live *H. pylori* significantly induced increased expression of CBS and CSE, signifying that *H. pylori* infection is associated with an increased generation of H_2S. **c** Detection of H_2S with gas chromatography with an inflammable detector. Significant amounts of H_2S were detected in gastric juice aspirated from a patient with *H. pylori*-associated chronic gastritis, signifying that H_2S might be responsible for *H. pylori*-associated gastric inflammation.

a relatively unexplored field of medicine except for CO poisoning and some anesthesiologic applications, a recent increase in the number of publications on medical gas therapies clearly indicates that there are significant opportunities for the use of gases as therapeutic tools for a variety of disease conditions or therapies. On the other hand, the other side of the coin of these medical gases always should be considered because H_2S are implicated principally in sepsis, organ damage, halitosis, and some potential risk factors for carcinogenesis. Conclusively, as the third gaseous mediator, H_2S may be a double-edged sword in the GI tract; it acts as an anti-inflammatory mediator to inhibit leukocyte activation, but may contribute to the inflammation in an already inflamed area when overproduced. It acts as a provoker or blocker of apoptosis and a pain provoker or reliever, alleviating the risk of NSAID injury through preserving perfusion but being a rascal for halitosis, leading to a rather confusing implication of H_2S in the GI tract. Therefore, the usefulness of H_2S gas in the GI tract is still debatable, but more effort should be made to elucidate a method to only pick out the beneficial aspects of the third gas of interest, H_2S.

Acknowledgement

The study received grants from the Ministry of Health and Family Welfare and the Ministry of Education and Scientific Technology.

References

1 Fiorucci S, Distrutti E, Cirino G, Wallace JL: The emerging roles of hydrogen sulfide in the gastrointestinal tract and liver. Gastroenterology 2006;131:259–271.

2 Mancardi D, Penna C, Merlino A, Del Soldato P, Wink DA, Pagliaro P: Physiological and pharmacological features of the novel gasotransmitter: hydrogen sulfide. Biochim Biophys Acta 2009;1787:864–872.

3 Kasparek MS, Linden DR, Kreis ME, Sarr MG: Gasotransmitters in the gastrointestinal tract. Surgery 2008;143:455–459.

4 Kashiba M, Kajimura M, Goda N, Suematsu M: From O_2 to H_2S: a landscape view of gas biology. Keio J Med 2002;51:1–10.

5 Wang R: Two's company, three's a crowd: can H_2S be the third endogenous gaseous transmitter? FASEB J 2002;16:1792–1798.

6 Abe K, Kimura H: The possible role of hydrogen sulfide as an endogenous neuromodulator. J Neurosci 1996;16:1066–1071.

7 Lou LX, Geng B, Du JB, Tang CS: Hydrogen sulphide-induced hypothermia attenuates stress-related ulceration in rats. Clin Exp Pharmacol Physiol 2008;35:223–228.

8 Nakao A, Sugimoto R, Billiar TR, McCurry KR: Therapeutic antioxidant medical gas. J Clin Biochem Nutr 2009;44:1–13.

9 Szabo C: Hydrogen sulphide and its therapeutic potential. Nat Rev Drug Discov 2007;6:917–935.

10 Moore PK, Bhatia M, Moochhala S: Hydrogen sulfide: from the smell of the past to the mediator of the future? Trends Pharmacol Sci 2003;24:609–611.

11 Medeiros JV, Bezerra VH, Gomes AS, Barbosa AL, Lima-Junior RC, Soares PM, Brito GA, Ribeiro RA, Cunha FQ, Souza MH: Hydrogen sulfide prevents ethanol-induced gastric damage in mice: role of ATP-sensitive potassium channels and capsaicin-sensitive primary afferent neurons. J Pharmacol Exp Ther 2009;330:764–770.

12 Wallace JL, Caliendo G, Santagada V, Cirino G, Fiorucci S: Gastrointestinal safety and anti-inflammatory effects of a hydrogen sulfide-releasing diclofenac derivative in the rat. Gastroenterology 2007;132:261–271.

13 Wallace JL: Hydrogen sulfide-releasing anti-inflammatory drugs. Trends Pharmacol Sci 2007;28:501–505.

14 Zanardo RC, Brancaleone V, Distrutti E, Fiorucci S, Cirino G, Wallace JL: Hydrogen sulfide is an endogenous modulator of leukocyte-mediated inflammation. FASEB J 2006;20:2118–2120.

15 Wallace JL, Dicay M, McKnight W, Martin GR: Hydrogen sulfide enhances ulcer healing in rats. FASEB J 2007;21:4070–4076.

16 Distrutti E, Sediari L, Mencarelli A, Renga B, Orlandi S, Antonelli E, Roviezzo F, Morelli A, Cirino G, Wallace JL, Fiorucci S: Evidence that hydrogen sulfide exerts antinociceptive effects in the gastrointestinal tract by activating KATP channels. J Pharmacol Exp Ther 2006;316:325–335.

17 Jha S, Calvert JW, Duranski MR, Ramachandran A, Lefer DJ: Hydrogen sulfide attenuates hepatic ischemia-reperfusion injury: role of antioxidant and antiapoptotic signaling. Am J Physiol Heart Circ Physiol 2008;295:H801–H806.

18 Yang G, Wu L, Wang R: Pro-apoptotic effect of endogenous H_2S on human aorta smooth muscle cells. FASEB J 2006;20:553–555.

19 Li L, Bhatia M, Zhu YZ, Zhu YC, Ramnath RD, Wang ZJ, Anuar FB, Whiteman M, Salto-Tellez M, Moore PK: Hydrogen sulfide is a novel mediator of lipopolysaccharide-induced inflammation in the mouse. FASEB J 2005;19:1196–1198.

20 Zhang H, Zhi L, Moochhala SM, Moore PK, Bhatia M: Endogenous hydrogen sulfide regulates leukocyte trafficking in cecal ligation and puncture-induced sepsis. J Leukoc Biol 2007;82:894–905.

21 Bhatia M, Wong FL, Fu D, Lau HY, Moochhala SM, Moore PK: Role of hydrogen sulfide in acute pancreatitis and associated lung injury. FASEB J 2005;19:623–625.

22 Yang G, Yang W, Wu L, Wang R: H_2S, endoplasmic reticulum stress, and apoptosis of insulin-secreting beta cells. J Biol Chem 2007;282:16567–16576.

23 Pae HO, Lee YC, Jo EK, Chung HT: Subtle interplay of endogenous bioactive gases (NO, CO and H(2)S) in inflammation. Arch Pharm Res 2009;32:1155–1162.

24 Lee SJ, Park JY, Choi KS, Ock CY, Hong KS, Kim YJ, Chung JW, Hahm KB: Efficacy of Korean red ginseng supplementation on eradication rate and gastric volatile sulfure compound levels after *Helicobacter pylori* eradication therapy. J Ginseng Res 2010;34:122–131.

25 Li L, Moore PK: An overview of the biological significance of endogenous gases: new roles for old molecules. Biochem Soc Trans 2007;35:1138–1141.

Prof. Ki-Baik Hahm
Laboratory of Translational Medicine, Lee Gil Ya Cancer and Diabetes Institute
Gachon University of Medicine and Science 7–45 Songdo-dong
Yeonsu-gu Incheon, 406–840 (Korea)
Tel. +82 32 899 6055, Fax +82 32 899 6054, E-Mail hahmkb@hotmail.com

Yoshikawa T, Naito Y (eds): Gas Biology Research in Clinical Practice.
Basel, Karger, 2011, pp 73–80

Role of Hydrogen Sulfide in Colitis

Tomohisa Takagi · Yuji Naito · Ikuhiro Hirata · Toshikazu Yoshikawa

Molecular Gastroenterology and Hepatology, Graduate School of Medical Science, Kyoto Prefectural University of Medicine, Kyoto, Japan

Abstract

Endogenous hydrogen sulfide (H_2S), which has been suggested to be a novel gas transmitter, is produced endogenously in mammalian tissue from L-cysteine by two enzymes, cystathionine-β-synthetase (CBS) and cystathionine-γ-lyase (CSE). Recent studies have shown that H_2S plays key roles in a number of biological processes, including inflammation, vasorelaxation, apoptosis, ischemia-reperfusion and oxidative stress. According to the results obtained from studies on several animal colitis models, H_2S has been implicated in the regulation of intestinal inflammation, and has gained importance as a potential therapeutic gas mediator in the management of intestinal inflammation. In this review, we present the current knowledge on the role of H_2S in colitis.

Inflammatory bowel diseases (IBDs) such as ulcerative colitis (UC) and Crohn's disease (CD) are chronic and recurrent intestinal inflammatory disorders. The incidence of IBD is rapidly increasing in Japan as well as in western countries [1]. While certain features of IBD suggest several possible causes, including familial, genetic, infectious, physiological, and immunological factors [2, 3], the precise pathogenesis of IBD remains unknown. Therefore, it is important to investigate the pathogenesis of IBD and to identify novel therapeutic molecules for IBDs.

Hydrogen sulfide (H_2S) is a colorless, water-soluble gas with the unpleasant smell of rotten eggs. Like nitric oxide (NO) and carbon monoxide (CO), H_2S has been recognized not only as a pollutant but also as an important gaseous physiological mediator [4]. H_2S is synthesized endogenously from L-cysteine in various mammalian tissues by two enzymes: cystathionine-β-synthetase (CBS) and cystathionine-γ-lyase (CSE). CBS is the predominant enzyme in the brain, nervous system, liver and kidney. CSE is expressed mainly in the liver and in both vascular and nonvascular smooth muscles. Low levels of CSE are also detectable in the gastrointestinal tract of rodents [5]. Regarding the physiological role of H_2S, it has been reported that H_2S mediates vasodilation and induces analgesia by opening K^{+}_{ATP} channels [5].

Furthermore, the involvement of H_2S in inflammation has been studied in various disease models. Accumulating evidence suggest that H_2S exerts anti-inflammatory effects at low concentrations [6–10], although some papers have also demonstrated pro-inflammatory effects of H_2S [11–16]. In this article, the roles of H_2S in gastrointestinal inflammation, especially in animal models of colitis, are reviewed and discussed.

H_2S Synthesis in the Gastrointestinal Tract

Methionine, an essential amino acid, is obtained by dietary intake and transported to the liver. Methionine adenosyltransferase (MAT) catalyzes the ATP-dependent conversion of methionine to S-adenosylmethionine. S-adenosylmethionine is converted to S-adenosylhomocysteine by the glycine N-methyltransferase (GNMT) and then hydrolyzed to homocysteine. Cystathionine-β-synthetase (CBS), which is regulated by S-adenosylmethionine, catalyzes the production of cystathionine by adding serine to homocysteine. Cystathionine can then be converted to cysteine via CSE. Cysteine, which can also be produced via reduction of dietary cystine, can then be converted to ammonium, pyruvate, and H_2S via the actions of either CSE or CBS. CBS is the predominant H_2S-generating enzyme in the brain and nervous system [17, 18], liver and kidney [19]. CSE is mainly expressed in the liver and in vascular and nonvascular smooth muscle.

In regard to the colonic expression of CBS and CSE, we found that the CBS and CSE mRNA expression levels and the H_2S contents in the colonic mucosa were increased with time throughout development of dextran sodium sulfate (DSS)-induced colitis in mice; a model of colitis induced in this manner is commonly used as a UC model (fig. 1a, b) [20]. In agreement with our data, Wallace et al. [21] demonstrated that colonic H_2S synthesis was markedly elevated after induction of 2,4,6-trinitrobenzenesulfonic acid (TNBS) colitis, which is used as a CD model, and then H_2S synthesis returned to the control levels when the colitis was resolved.

Schicho et al. [22] demonstrated the expression of CSE and CBS in the enteric nervous system of the guinea pig and human intestine. Martin et al. [23] demonstrated that H_2S was synthesized throughout the gastrointestinal tract, and CSE and CBS were constitutively expressed in these tissues in both rats and mice, as well as in the healthy human colon. They also described that CBS appears to be the predominant enzymatic source of H_2S in the colon, but not in the other gastrointestinal tissues, in rat. It has been shown that CSE is expressed in the liver, stomach, small intestine and colon of mice [7, 24]. Immunoreactive CSE and CBS have been identified in over 90% of the submucous and myenteric neurons in the human and guinea pig colon [22], as well as in colonic smooth muscle in mice. In the rat, CSE expression is most evident in the proximal regions of the gastrointestinal tract (stomach, duodenum, jejunum), and occurs at comparatively low levels in the ileum and colon. Conversely, the levels

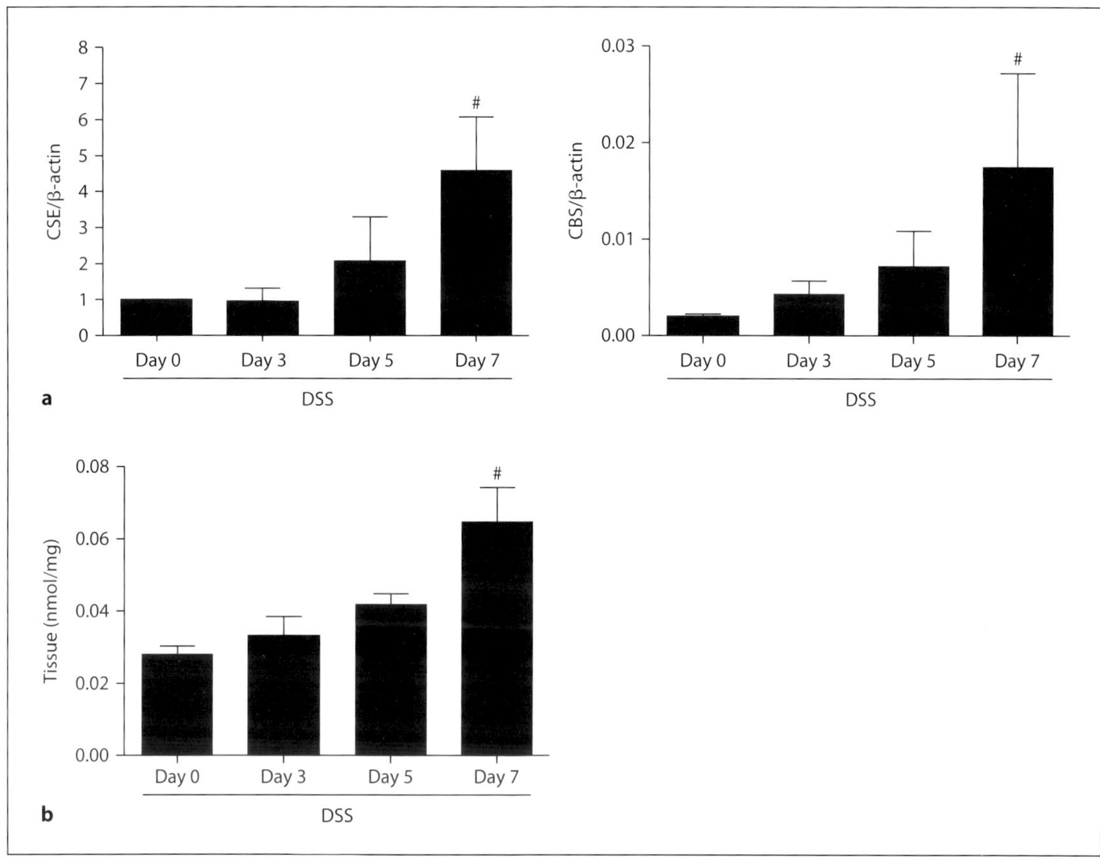

Fig. 1. Chronological relative expression of the mRNAs for CSE and CBS (**a**) and the amounts of H_2S (**b**) in the colonic mucosa of mice administered 8% DSS. Each value indicates the mean ± SEM for 5–7 mice. # $p < 0.05$ when compared to day 0 [20].

of CBS expression were relatively low in the duodenum and jejunum, but high in the stomach, ileum and colon.

Role of H_2S in colonic inflammation

Effect of the Inhibitors/Donors of H_2S on Colitis Severity

Whether the elevated H_2S synthesis in the inflamed colon contributes to the disease activity of colitis was investigated by treating animals with experimental colitis with a number of inhibitors of H_2S synthesis, including an inhibitor of CBS (CHH), a reversible inhibitor of CSE (BCA), and an irreversible inhibitor of CSE (PAG).

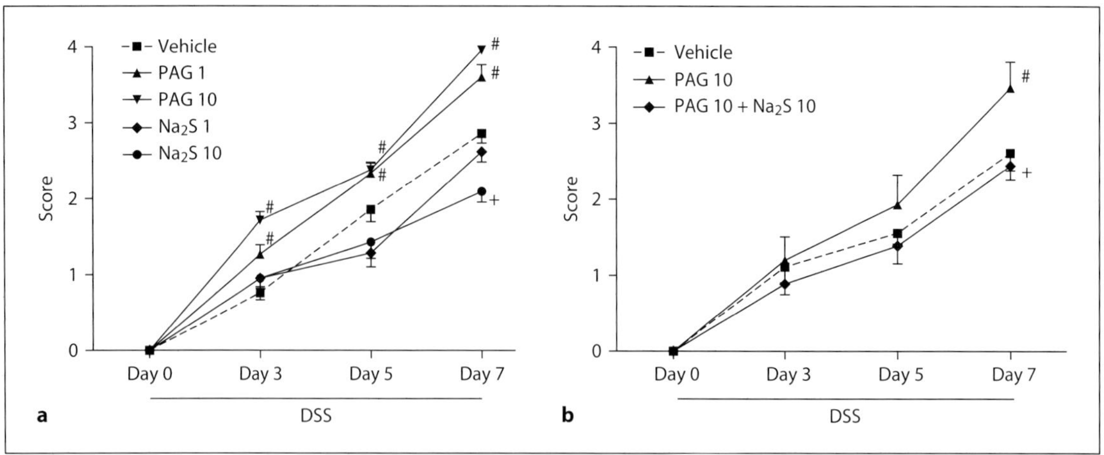

Fig. 2. The disease activity index (DAI) was determined by scoring changes in body weight, stool consistency and blood in the stool. **a** The effect of PAG or Na_2S on DAI scores. Each value indicates the mean ± SEM for 7 mice. $^{\#,+}$ $p < 0.05$ when compared to the vehicle-treated group. **b** The effect of PAG with or without Na_2S on DAI scores. Each value indicates the mean ± SEM for 7 mice. $^{\#}$ $p < 0.05$ when compared to the vehicle-treated group, $^{+}$ $p < 0.05$ when compared to the PAG-treated group [20].

In our study, inhibiting the endogenous H_2S-generating enzyme by PAG significantly deteriorated DSS-induced colitis in mice. On the other hand, the administration of Na_2S, an H_2S donor, ameliorated DSS-induced colitis. Moreover, Na_2S significantly negated the effect of PAG (fig. 2a, b) [20].

Wallace et al. [21] also demonstrated that endogenous and exogenous H_2S promotes resolution of colitis in rats. Inhibiting H_2S synthesis by CHH or BCA significantly exacerbated the TNBS model of colitis in rats. On the other hand, administration of an H_2S donor (NaHS) resulted in a significant reduction in the severity of colitis.

Interestingly, a study on an H_2S-releasing derivative of mesalamine (ATB-429; Antibe Therapeutics Inc., Toronto, Ont., Canada) by Fiorucci et al. [25] supported the notion that H_2S exerts beneficial effects in the context of colitis. In TNBS-induced colitis in mice, treatment with ATB-429 twice daily for 1 week resulted in significantly more-rapid resolution of colitis than did treatment with mesalamine.

H_2S Modulates Leukocyte/Neutrophil Behavior

Many investigators, including us, have hypothesized that neutrophil-mediated inflammation is involved in the development of DSS-induced colonic mucosal injury. Three lines of evidence support this hypothesis: (1) colonic mucosal endothelial intercellular adhesion molecule 1 (ICAM-1) expression is enhanced at an early stage in the

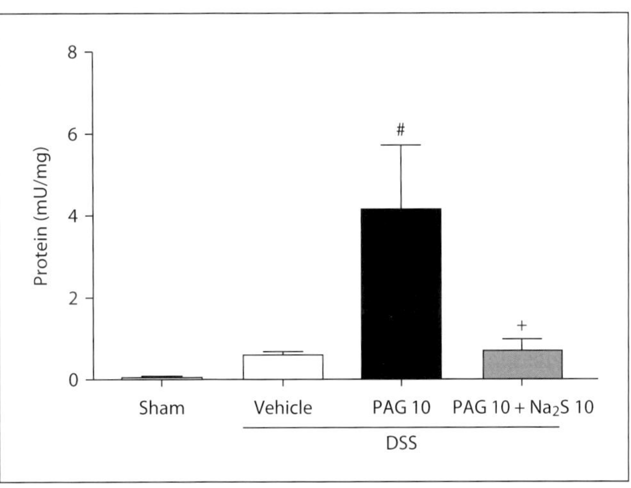

Fig. 3. Effect of PAG with or without Na_2S on neutrophil accumulation expressed as myeloperoxidase (MPO) activity in the colonic mucosa of mice administered 8% DSS. Each value indicates the mean ± SEM for 5–7 mice. $^{\#}$ $p < 0.05$ when compared to the vehicle-treated group, $^{+}$ $p < 0.05$ when compared to the PAG-treated group [20].

inflammatory cascade of DSS-induced colitis [26]; (2) selective depletion of neutrophils by monoclonal antibody RP-3 suppresses colitis in rats [27], and (3) immunoneutralization of ICAM-1 on endothelial cells significantly attenuates colonic mucosal injury and neutrophil accumulation in rats [28]. We showed that myeloperoxidase (MPO) activity, an index of tissue-associated neutrophil accumulation, significantly increased in the colonic mucosa by inhibiting endogenous H_2S generation, and this increase was significantly inhibited by co-treatment with an H_2S donor. These results indicate that the inhibition of neutrophil accumulation by H_2S may be one of the protective factors helping to decrease DSS-induced colitis (fig. 3). The inhibitory effect of H_2S against neutrophil accumulation is also supported by previous reports, in which H_2S inhibited neutrophil-endothelial interaction. Zanardo et al. [8] demonstrated that H_2S inhibited neutrophil adhesion/activation in carrageenan-induced paw edema. The induction of adhesive interactions between circulating leukocytes and the vascular endothelium is the initial step in the process of extravasation of leukocytes and their migration to sites of injury or infection. An important role of H_2S as a tonic regulator of this process was suggested by the observation that inhibition of H_2S synthesis in rats resulted in a rapid increase in leukocyte adherence in mesenteric venules [8]. When H_2S synthesis is inhibited, an increase in the expression of adhesion molecules on the endothelium (P-selectin, ICAM-1) and on leukocytes (LFA-1) can be detected [7]. Altered leukocyte interactions with the endothelium were observed in mice genetically deficient in one of the key enzymes for H_2S synthesis (CBS). Mice completely deficient in this enzyme did not survive, but those that were heterozygous for CBS exhibited increased vascular permeability, reduced leukocyte-rolling velocity, and increased leukocyte adherence to the vascular endothelium [29]. Fiorucci et al. [7] showed that H_2S suppresses nonsteroidal anti-inflammatory drug-induced gastropathy by suppressing leukocyte adherence via the opening of K^+_{ATP} channels.

Spiller et al. [30] also demonstrated that murine sepsis induced by cecal ligation and puncture was inhibited by pretreatments with H_2S donors (NaHS) according to the inhibition of neutrophil migration. As this beneficial effect of H_2S were blocked by glibenclamide (an ATP-dependent K^+ channel blocker), the anti-inflammatory actions of H_2S may be mediated through activation of K^+_{ATP} channels.

Other Possible Functions of H_2S in Colitis

Several lines of evidence have shown that H_2S downregulates the expression of pro-inflammatory mediators. H_2S donors reduced tumor necrosis factor (TNF)-α mRNA and protein levels in the colon in a rat model of colitis, in association with a significant reduction in the severity of the colitis [21]. An H_2S-releasing derivative of mesalamine (ATB-429) reduced the expression of mRNA for several key pro-inflammatory cytokines/chemokines (e.g. TNF-α, interferon (IFN)-γ) in TNBS-induced colitis in mice [25].

Our recent studies have shown that inhibiting H_2S with PAG significantly enhanced lipid peroxidation in DSS-induced inflamed colonic mucosa, and lipid peroxidation was significantly inhibited by supplementing H_2S with Na_2S. Polyunsaturated fatty acids of cell membranes are degraded by lipid peroxidation, with subsequent disruption of membrane integrity, suggesting that lipid peroxidation mediated by oxygen radicals is an important cause of damage to and destruction of cell membranes [31, 32]. Induction of lipid peroxidation is a critical early event in this experimental inflammatory bowel disease model [33]. Some previous reports have demonstrated the antioxidant and cytoprotective effects of H_2S, which may be related to the neutralization of various reactive species, including oxyradicals [34], peroxynitrite [35], and hypochlorous acid [36]. H_2S can also upregulate the endogenous antioxidant system [37].

Future Perspectives

Novel therapeutic targets are required to control inflammatory bowel disease because of its complicated pathogenesis. Recent research has increasingly focused attention on gaseous mediators, including not only NO, CO but also H_2S. In the present article, we reviewed recent findings on the potential anti-inflammatory actions of H_2S in experimental colitis models. It has been suggested for a number of years that high levels of H_2S produced by bacteria could contribute to colitis and possibly to colon cancer. However, these theories have been controversial, because most of the H_2S that is produced in the lumen of the intestine is absorbed and rapidly metabolized. Furne et al. [38] showed that treatment with dextran sulfate did not increase the release of H_2S in feces, and scavenging excessive intraluminal H_2S using bismuth did not affect the dextran sulfate-induced colitis. At any rate, further research is needed to clarify the molecular mechanisms underlying the cellular signaling of sulfide and the interactions between H_2S and NO/CO.

References

1 Asakura K, Nishiwaki Y, Inoue N, Hibi T, Watanabe M, Takebayashi T: Prevalence of ulcerative colitis and Crohn's disease in japan. J Gastroenterol 2009; 44:659–665.

2 Podolsky DK: Inflammatory bowel disease. N Engl J Med 2002;347:417–429.

3 Xavier RJ, Podolsky DK: Unravelling the pathogenesis of inflammatory bowel disease. Nature 2007;448: 427–434.

4 Szabo C: Hydrogen sulphide and its therapeutic potential. Nat Rev Drug Discovery 2007;6:917–935.

5 Fiorucci S, Distrutti E, Cirino G, Wallace JL: The emerging roles of hydrogen sulfide in the gastrointestinal tract and liver. Gastroenterology 2006;131: 259–271.

6 Johansen D, Ytrehus K, Baxter GF: Exogenous hydrogen sulfide (H_2S) protects against regional myocardial ischemia-reperfusion injury: evidence for a role of katp channels. Basic Res Cardiol 2006; 101:53–60.

7 Fiorucci S, Antonelli E, Distrutti E, Rizzo G, Mencarelli A, Orlandi S, Zanardo R, Renga B, Di Sante M, Morelli A, Cirino G, Wallace JL: Inhibition of hydrogen sulfide generation contributes to gastric injury caused by anti-inflammatory nonsteroidal drugs. Gastroenterology 2005;129:1210–1224.

8 Zanardo RC, Brancaleone V, Distrutti E, Fiorucci S, Cirino G, Wallace JL: Hydrogen sulfide is an endogenous modulator of leukocyte-mediated inflammation. FASEB J 2006;20:2118–2120.

9 Distrutti E, Sediari L, Mencarelli A, Renga B, Orlandi S, Antonelli E, Roviezzo F, Morelli A, Cirino G, Wallace JL, Fiorucci S: Evidence that hydrogen sulfide exerts antinociceptive effects in the gastrointestinal tract by activating katp channels. J Pharmacol Exp Ther 2006;316:325–335.

10 Distrutti E, Sediari L, Mencarelli A, Renga B, Orlandi S, Russo G, Caliendo G, Santagada V, Cirino G, Wallace JL, Fiorucci S: 5-Amino-2- hydroxybenzoic acid 4-(5-thioxo-5h-[1,2]dithiol-3yl)-phenyl ester (atb-429), a hydrogen sulfide-releasing derivative of mesalamine, exerts antinociceptive effects in a model of postinflammatory hypersensitivity. J Pharmacol Exp Ther 2006;319:447–458.

11 Qu K, Chen CPLH, Halliwell B, Moore PK, Wong PTH: Hydrogen sulfide is a mediator of cerebral ischemic damage. Stroke 2006;37:889–893.

12 Bhatia M, Wong FL, Fu D, Lau HY, Moochhala SM, Moore PK: Role of hydrogen sulfide in acute pancreatitis and associated lung injury. FASEB J 2005;19: 623–625.

13 Li L, Bhatia M, Zhu YZ, Zhu YC, Ramnath RD, Wang ZJ, Anuar FBM, Whiteman M, Salto-Tellez M, Moore PK: Hydrogen sulfide is a novel mediator of lipopolysaccharide-induced inflammation in the mouse. FASEB J 2005;19:1196–1198.

14 Collin M, Anuar FBM, Murch O, Bhatia M, Moore PK, Thiemermann C: Inhibition of endogenous hydrogen sulfide formation reduces the organ injury caused by endotoxemia. Br J Pharmacol 2005;146: 498–505.

15 Bhatia M, Sidhapuriwala J, Moochhala SM, Moore PK: Hydrogen sulphide is a mediator of carrageenan-induced hindpaw oedema in the rat. Br J Pharmacol 2005;145:141–144.

16 Zhang H, Zhi L, Moore PK, Bhatia M: Role of hydrogen sulfide in cecal ligation and puncture-induced sepsis in the mouse. Am J Physiol Lung Cell Mol Physiol 2006;290:1193–1201.

17 Abe K, Kimura H: The possible role of hydrogen sulfide as an endogenous neuromodulator. J Neurosci 1996;16:1066–1071.

18 Kimura H: Hydrogen sulfide: from brain to gut. Antioxidants Redox Signaling 2010;12:1111–1123.

19 Wang R: Two's company, three's a crowd: can H_2S be the third endogenous gaseous transmitter? FASEB J 2002;16:1792–1798.

20 Hirata I, Naito Y, Takagi T, Mizushima K, Suzuki T, Omatsu T, Handa O, Ichikawa H, Ueda H, Yoshikawa T: Endogenous hydrogen sulfide is an anti-inflammatory molecule in dextran sodium sulfate-induced colitis in mice. Dig Dis Sci (in press).

21 Wallace JL, Vong L, McKnight W, Dicay M, Martin GR: Endogenous and exogenous hydrogen sulfide promotes resolution of colitis in rats. Gastroenterology 2009;137:569–578, e561.

22 Schicho R, Krueger D, Zeller F, Von Weyhern CW, Frieling T, Kimura H, Ishii I, De Giorgio R, Campi B, Schemann M: Hydrogen sulfide is a novel prosecretory neuromodulator in the guinea-pig and human colon. Gastroenterology 2006;131:1542–1552.

23 Martin GR, McKnight GW, Dicay MS, Coffin CS, Ferraz JGP, Wallace JL: Hydrogen sulphide synthesis in the rat and mouse gastrointestinal tract. Dig Liver Dis 2010;42:103–109.

24 Linden DR, Sha L, Mazzone A, Stoltz GJ, Bernard CE, Furne JK, Levitt MD, Farrugia G, Szurszewski JH: Production of the gaseous signal molecule hydrogen sulfide in mouse tissues. J Neurochem 2008;106:1577–1585.

25 Fiorucci S, Orlandi S, Mencarelli A, Caliendo G, Santagada V, Distrutti E, Santucci L, Cirino G, Wallace JL: Enhanced activity of a hydrogen sulphide-releasing derivative of mesalamine (ATB-429) in a mouse model of colitis. Br J Pharmacol 2007;150:996–1002.

26 Breider MA, Eppinger M, Gough A: Intercellular adhesion molecule-1 expression in dextran sodium sulfate-induced colitis in rats. Vet Pathol 1997;34: 598–604.

27 Natsui M, Kawasaki K, Takizawa H, Hayashi SI, Matsuda Y, Sugimura K, Seki K, Narisawa R, Sendo F, Asakura H: Selective depletion of neutrophils by a monoclonal antibody, RP-3, suppresses dextran sulphate sodium-induced colitis in rats. J Gastroenterol Hepatol 1997;12:801–808.

28 Taniguchi T, Tsukada H, Nakamura H, Kodama M, Fukuda K, Saito T, Miyasaka M, Seino Y: Effects of the anti-ICAM-1 monoclonal antibody on dextran sodium sulphate-induced colitis in rats. J Gastroenterol Hepatol 1998;13:945–949.

29 Kamath AF, Chauhan AK, Kisucka J, Dole VS, Loscalzo J, Handy DE, Wagner DD: Elevated levels of homocysteine compromise blood-brain barrier integrity in mice. Blood 2006;107:591–593.

30 Spiller F, Orrico MI, Nascimento DC, Czaikoski PG, Souto FO, Alves-Filho JC, Freitas A, Carlos D, Montenegro MF, Neto AF, Ferreira SH, Rossi MA, Hothersall JS, Assreuy J, Cunha FQ: Hydrogen sulfide improves neutrophil migration and survival in sepsis via K+ATP channel activation. Am J Respir Crit Care Med 2010;182:360–368.

31 Niki E, Komuro E: Inhibition of peroxidation of membranes. Basic Life Sci 1988;49:561–566.

32 Niki E, Noguchi N, Gotoh N: Dynamics of lipid peroxidation and its inhibition by antioxidants. Biochem Soc Trans 1993;21:313–317.

33 Naito Y, Takagi T, Yoshikawa T: Neutrophil-dependent oxidative stress in ulcerative colitis. J Clin Biochem Nutr 2007;41:18–26.

34 Geng B, Chang L, Pan C, Qi Y, Zhao J, Pang Y, Du J, Tang C: Endogenous hydrogen sulfide regulation of myocardial injury induced by isoproterenol. Biochem Biophys Res Commun 2004;318:756–763.

35 Whiteman M, Armstrong JS, Chu SH, Jia-Ling S, Wong BS, Cheung NS, Halliwell B, Moore PK: The novel neuromodulator hydrogen sulfide: An endogenous peroxynitrite 'scavenger'? J Neurochem 2004; 90:765–768.

36 Whiteman M, Cheung NS, Zhu YZ, Chu SH, Siau JL, Wong BS, Armstrong JS, Moore PK: Hydrogen sulphide: a novel inhibitor of hypochlorous acid-mediated oxidative damage in the brain? Biochem Biophys Res Commun 2005;326:794–798.

37 Kimura Y, Kimura H: Hydrogen sulfide protects neurons from oxidative stress. FASEB J 2004;18: 1165–1167.

38 Furne JK, Suarez FL, Ewing SL, Springfield J, Levitt MD: Binding of hydrogen sulfide by bismuth does not prevent dextran sulfate-induced colitis in rats. Dig Dis Sci 2000;45:1439–1443.

Yuji Naito, MD, PhD
Molecular Gastroenterology and Hepatology, Graduate School of Medical Science
Kyoto Prefectural University of Medicine
465 Kajii-cho, Kamigyo-ku Kyoto 602-8566 (Japan)
Tel. +81 75 251 5519, Fax +81 75 251 0710, E-Mail ynaito@koto.kpu-m.ac.jp

Yoshikawa T, Naito Y (eds): Gas Biology Research in Clinical Practice.
Basel, Karger, 2011, pp 81–90

HCO_3^- Stimulatory Action of Hydrogen Sulfide in Rat Duodenum

Koji Takeuchi · Fumitaka Ise · Masafumi Koyama · Koji Dogishi · Masashi Yasuda · Shusaku Hayashi

Division of Pathological Sciences Department of Pharmacology and Experimental Therapeutics
Kyoto Pharmaceutical University Misasagi, Yamashina, Kyoto, Japan

Abstract

Hydrogen sulfide (H_2S), generated in the mammalian body during the metabolism of cysteine, has been recognized as a gaseous signaling molecule in addition to nitric oxide (NO) and carbon monoxide. We examined the effect of H_2S on duodenal HCO_3^- secretion in rats and investigated the mechanism involved in this response. Animals were fasted for 18 h and anesthetized with urethane. A duodenal loop was perfused with saline, and HCO_3^- secretion was measured at pH 7.0 using a pH stat method. The loop was perfused at a rate of 0.2 ml/min with NaHS (H_2S donor) for 5 min or 10 mM HCl for 10 min. Indomethacin or L-NAME was given s.c. 30 min or 3 h, respectively, before NaHS or acidification, while glibenclamide (K_{ATP} channel blocker) or propargylglycine (cystathionine-γ-lyase inhibitor) was given i.p. 30 min before. Mucosal perfusion with NaHS dose-dependently increased the HCO_3^- secretion, and this effect was significantly attenuated by indomethacin, L-NAME and sensory deafferentation but not by glibenclamide. Mucosal PGE_2 production and luminal release of NO were both increased by NaHS perfusion. Mucosal acidification stimulated HCO_3^- secretion concomitant with increase in PGE_2 and NO production, and these responses were mitigated by propargylglycine. We conclude that H_2S increases HCO_3^- secretion in the rat duodenum, and that this action is partly mediated by PG and NO as well as by capsaicin-sensitive afferent neurons. It is assumed that endogenous H_2S is involved in the regulatory mechanism of acid-induced HCO_3^- secretion and mucosal protection in the duodenum.

Hydrogen sulfide (H_2S), which is a toxic gas and common industrial pollutant, is found endogenously in the mammalian body where it is generated during the metabolism of cysteine. Recently, H_2S has been recognized as a gaseous signaling molecule, in addition to NO and carbon monoxide (CO), which are both known to act as endogenous vasodilator, neuromodulators, and anti-inflammatory mediators [1–4]. In the gastrointestinal tract, H_2S plays a role in gastric mucosal protection and the resolution of colitis as well as the regulation of intestinal secretion [5–8]. In addition, the finding that gastrointestinal tissues are capable of producing H_2S has led to the hypothesis that H_2S is an endogenous gaseous signaling molecule [9, 10]. Two enzymes, cystathionine β-synthase (CBS) and cystathionine γ-lyase (CSE), are responsible for generation of H_2S from the amino acid L-cysteine in mammalian tissues [2], and they are widely expressed in the gastrointestinal tract [4]. NaHS, an H_2S donor, has previously been reported to increase gastric mucosal blood flow via K^+_{ATP} channel and to attenuate gastric mucosal injury induced by nonsteroidal anti-inflammatory drugs (NSAIDs) [11]. These findings suggest that H_2S plays a role in the physiology or pathophysiology of the gastrointestinal tract.

We herein introduced our data on the stimulatory effect of an H_2S donor, NaHS, on HCO_3^- secretion in the rat duodenum and also demonstrated the involvement of endogenous H_2S in the regulation of the HCO_3^- response to mucosal acidification [12].

Materials and Methods

Male Sprague-Dawley rats (220–260 g, Nippon Charles River, Kanagawa, Japan) were used under urethane anesthesia. A duodenal loop (17 mm) was made between the pyloric ring and the area just above the outlet of the common bile duct to exclude the influences of bile and pancreatic juice [13]. The loop was then perfused a rate of 0.2 ml/min with saline that was gassed with 100% O_2, heated at 37°C and kept in a reservoir, and HCO_3^- secretion was measured at pH 7.0 using a pH-stat method (Hiranuma Comtite-8, Mito, Japan) with addition of 2 mM HCl to the reservoir. After basal HCO_3^- secretion was well stabilized, the duodenal loop was perfused at a rate of 0.2 ml/min with NaHS solution (0.1–1 mM) for 5 min or with 10 mM HCl for 10 min (mucosal acidification), rinsed with saline, and again perfused with saline. In some cases, the effects of indomethacin (a cyclooxygenase inhibitor), L-NAME (an NO synthase inhibitor), glibenclamide (a K^+_{ATP} channel inhibitor), and sensory deafferentation on the HCO_3^- response to NaHS were examined. Indomethacin (5 mg/kg) or L-NAME (20 mg/kg) was given s.c. 30 min or 3 h, respectively, before mucosal exposure to NaHS, while glibenclamide (10 mg/kg) or propargylglycine (10 mg/kg: a CSE antagonist) was given i.p. 30 min before NaHS [13–15]. Chemical ablation of capsaicin-sensitive afferent neurons was achieved with 3 consecutive s.c. injections of capsaicin (total dose, 100 mg/kg) once daily 2 weeks before the experiment [13, 16].

The acid-induced mucosal damage was induced by perfusion of the duodenal loop with 100 mM HCl at a flow rate of 1 ml/h for 4 h [15]. Indomethacin (5 mg/kg) or L-NAME (20 mg/kg) was given s.c. 1 h or 3 h, respectively, before acid perfusion, while propargylglycine (10 mg/kg) was given i.p. 30 min before the acidification. The area (mm^2) of each hemorrhagic lesion was measured, summed per duodenum, and expressed as a damage score.

The drugs used were urethane (Tokyo Kasei, Tokyo, Japan), indomethacin, L-NAME, glibenclamide, DL-propargylglycine, capsaicin (Sigma Chemicals, St. Louis, Mo., USA), terbutaline (Buricanyl®, Fujisawa, Osaka, Japan) and aminophylline (Neophylline®, Eizai, Tokyo, Japan). Indomethacin was suspended in saline with a drop of Tween 80 (Wako, Osaka, Japan). Capsaicin was dissolved in a Tween 80-ethanol solution (10% ethanol, 10% Tween and 80% saline, w/w: Wako) for s.c. injection, while it was suspended in a 0.5% carboxymethylcellulose solution (CMC: Nacalai Tesque) for topical application. Other agents were dissolved in saline. Each agent was prepared immediately before use and administered i.p. or s.c. in a volume of 0.5 ml/100 g body weight, or perfused in the duodenal loop at a rate of 0.2 ml per min with an infusion pump (Terumo, Tokyo, Japan). Control animals received saline in place of the active agents.

Data are presented as the mean ± SE from 4–6 rats per group. Statistical analyses were performed using a two-tailed unpaired t test and Dunnett's multiple comparison test, with $p < 0.05$ regarded as significant.

Results

Effect of NaHS on Duodenal HCO_3^- Secretion

Under urethane anesthesia, rat duodenum spontaneously secreted HCO_3^- at a steady rate of 0.3–0.6 μEq/10 min. Perfusion of the loop with NaHS (0.1–1 mM) for 5 min increased the secretion of HCO_3^- in a concentration-dependent manner, and this effect was significant at concentrations of 0.3 mM or more when compared with value obtained in the saline-perfused duodenum (fig. 1a). The HCO_3^- stimulatory effect of NaHS (0.5 mM) was significantly attenuated by prior administration of indomethacin and L-NAME as well as by chemical ablation of capsaicin-sensitive afferent neurons, although glibenclamide, an inhibitor of K^+_{ATP} channel, did not affect the HCO_3^- response induced by NaHS. In addition, NaHS caused a marked increase in HCO_3^- secretion even when endogenous H_2S production was inhibited by pretreatment of propargylglycine.

Perfusion of the duodenum with 0.5 mM NaHS for 5 min significantly increased the mucosal PGE_2 content as well as the luminal NO levels (fig. 1b). The increased generation of PGE_2 in response to NaHS treatment was significantly suppressed by prior administration of L-NAME as well as by indomethacin, while glibenclamide had no effect. By contrast, the increase of luminal NO release caused by NaHS was significantly abrogated by L-NAME, while neither indomethacin nor glibenclamide had any effect (fig. 1c).

Effect of Propargylglycine on Acid-Induced Duodenal HCO_3^- Secretion

Secretion of HCO_3^- in the duodenum was markedly increased when the mucosa was acidified by perfusion with 10 mM HCl for 10 min; the ΔHCO_3^- output was 3.5 ± 0.4 μEq/h, which was roughly 10 times greater than that observed in saline-perfused duodenum. The acid-induced HCO_3^- secretion was significantly attenuated by both indomethacin and L-NAME, which inhibited secretion by 76.5 and 67.6%, respectively (fig. 2a). Likewise, the response was also significantly inhibited (65.7%) by pretreatment of the animals with propargylglycine, a CSE inhibitor, administered; i.p. 30 min before acidification the ΔHCO_3^- output was 1.2 ± 0.4 μEq/h. However, glibenclamide had no effect on the response to mucosal acidification.

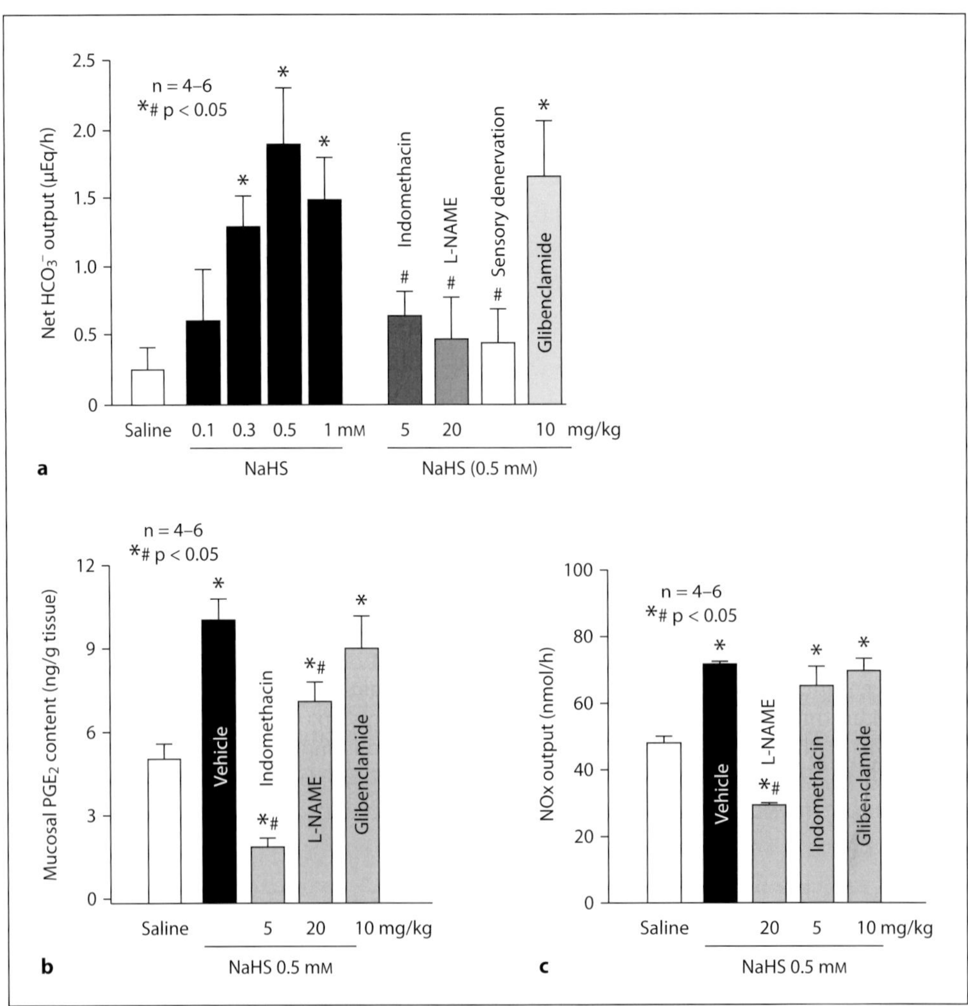

Fig. 1. Effects of NaHS on HCO_3^- secretion (**a**), PGE_2 content (**b**) and NO release (**c**) in rat duodenum in the absence or presence of various inhibitors. NaHS (0.5 mM) was locally applied to the mucosa by perfusion of the duodenal loop for 5 min. Indomethacin (5 mg/kg) or L-NAME (20 mg/kg) was given s.c. 30 min or 3 h before the treatment with NaHS, respectively, while glibenclamide (10 mg/kg) was given i.p. 30 min prior. Sensory deafferentation was achieved with 3 consecutive injections of capsaicin (total dose, 100 mg/kg) 2 weeks before the experiment. **a** Total ΔHCO_3^- output 1 h after treatment with NaHS and represent the mean ± SE from 4–6 rats. **b, c** Mean ± SE from 4–6 rats. Significant difference at $p < 0.05$ * vs. saline and # vs. vehicle. Data adopted after modification from Ise et al. [12].

Mucosal acidification significantly increased mucosal PGE_2 content and luminal NO release in the duodenum, in both cases approximately 2 times the level in saline-perfused rats. The enhanced PGE_2 generation was significantly inhibited by the prior administration of indomethacin or L-NAME (fig. 2b). By contrast, the acid-induced

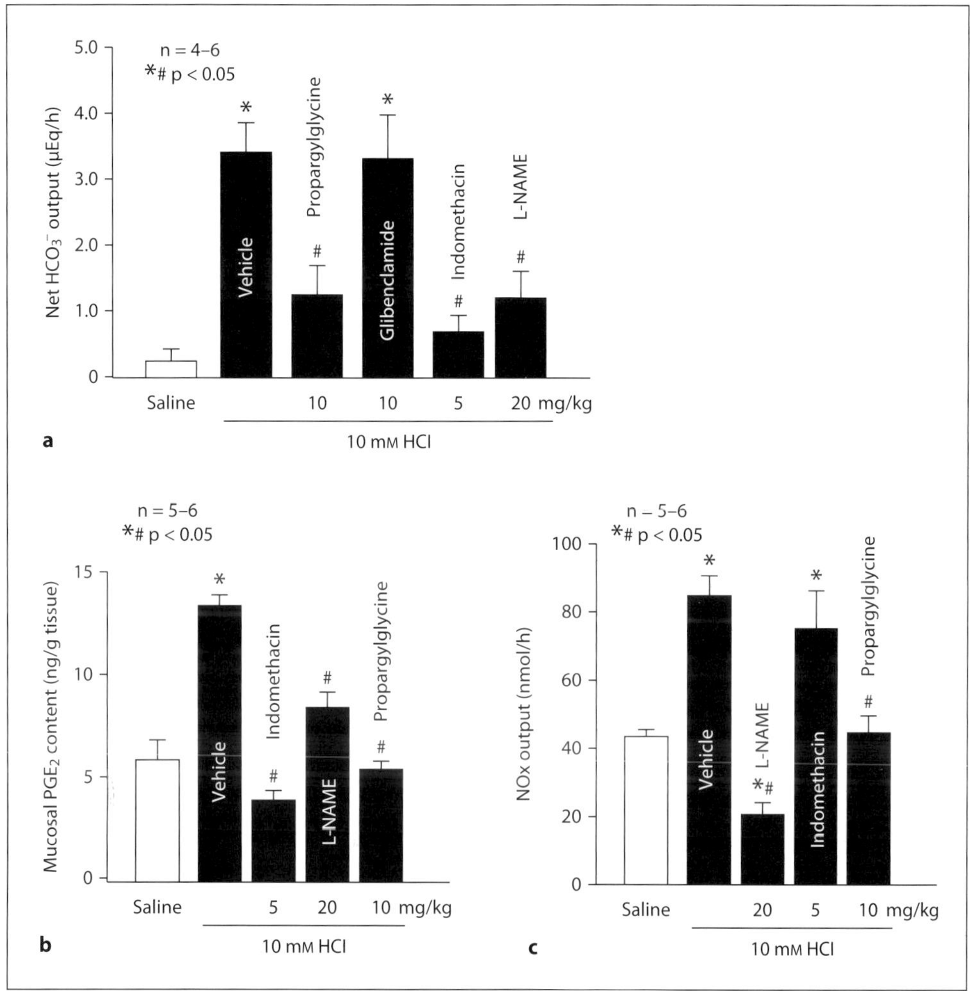

Fig. 2. Effects of propargylglycine, glibenclamide, indomethacin or L-NAME on HCO_3^- response (**a**), PGE_2 content (**b**) and NO release (**c**) induced by mucosal acidification in the rat duodenum. The acidification was performed by perfusion of the duodenal loop with 10 mM HCl for 10 min. Propargylglycine (10 mg/kg) or glibenclamide (10 mg/kg) was given i.p. 30 min before mucosal acidification, while indomethacin (5 mg/kg) or L-NAME (20 mg/kg) was given s.c. 30 min or 3 h, respectively, before acidification. **a** Total ΔHCO_3^- output 1 h after mucosal acidification and are presented as the means ± SE for 4–6 rats. **b** Mean ± SE for 5–6 rats. **c** Mean ± SE of values obtained for 1 h after acidification for 4–6 rats. Significant difference at $p < 0.05$ * vs. saline and # vs. vehicle. Data adopted after modification from Ise et al. [12].

NO release was totally attenuated by L-NAME but not affected by indomethacin (fig. 2c). On the other hand, propagylglycine totally suppressed both the increase of PGE_2 generation and NO release in the duodenum in response to acidification, the inhibition being 98.8 and 90.2%, respectively.

Effect of Propargylglycine on Acid-Induced Duodenal Damage

Perfusion of the proximal duodenum with 100 mM HCl for 4 h caused superficial, hemorrhagic damage in the mucosa, with a lesion score of 19.2 ± 0.9 mm^2. The lesion score was markedly aggravated in animals pretreated with indomethacin or L-NAME, the value being 48.3 ± 4.1 or 44.5 ± 5.1 mm^2, respectively. Likewise, when animals were pretreated with propargylglycine, a CSE antagonist, prior to the onset of acid perfusion, the severity of duodenal lesions was significantly worsened, the lesion score being 40.1 ± 4.6 mm^2.

Discussion

Since H_2S has been recognized as an important endogenous mediator, in addition to NO and CO, an accumulating body of evidence supports the role of this molecule in various aspects of gastrointestinal physiology, such as the maintenance of mucosal integrity, regulation of fluid transport and modulation of inflammation [5, 7, 8, 17, 18]. The present study showed for the first time that H_2S, generated exogenously from NaHS, stimulated the secretion of HCO_3^- in the duodenum, and this action was mediated by endogenous PGs and NO as well as by capsaicin-sensitive afferent neurons. We further demonstrated the involvement of endogenous H_2S in the regulation of acid-induced HCO_3^- secretion in the duodenum.

In the present study, we perfused the loop with NaHS (0.1–1 mM) at a rate of 0.2 ml/min for 5 min to locally generate H_2S in the duodenum. The levels of H_2S in human blood or the rat ileum have been reported to be 1.2 μmol/l or 6.5 nmol/min/g tissue [5]. Although we did not measure actual amount of H_2S generated in the duodenum, the amount of H_2S generated by NaHS treatment may not be much high compared to that generated endogenously. Indeed, the same doses of NaHS have been used to generate H_2S in various in vivo experiments by others [7, 11]. We found that NaHS increased duodenal HCO_3^- secretion in a concentration-dependent manner. Of interest, the HCO_3^- response to NaHS was significantly attenuated in the presence of indomethacin, suggesting the involvement of endogenous PGs in the stimulatory action of H_2S.

It has been shown that PGE_2 stimulates HCO_3^- secretion in the duodenum via activation of EP3/EP4 receptors [19, 21]. We observed that the stimulatory effect of NaHS on HCO_3^- secretion was significantly abrogated by pretreatment with selective EP3 and EP4 antagonists (data not shown), supporting the involvement of PGs in this action. Hu et al. [22] showed that H_2S caused upregulation of COX-2 expression and PGE_2 production in cardiac myocytes. Wallace et al. [8] reported that inhibition of H_2S production resulted in downregulation of COX-2 expression and PGE_2 synthesis in the colon. We also observed that local application of NaHS increased mucosal PGE_2 production in the duodenum. These results together suggest that H_2S

increases PGE_2 generation in the duodenum and stimulates HCO_3^- secretion mediated by endogenous PGE_2.

The HCO_3^- response induced by local application of NaHS was also significantly attenuated by L-NAME, a nonselective NO synthase inhibitor, suggesting the participation of endogenous NO in this process, in addition to PGs. Zhao et al. [23] reported that H_2S increased the release of NO from vascular endothelium. The present study showed an increase of luminal NO release in the duodenum after mucosal application of NaHS. On the other hand, we previously reported that NOR3, an NO donor, stimulated HCO_3^- secretion in the rat duodenum with concomitant increase of mucosal PGE_2 content [15]. Since the stimulatory effect of NOR3 was attenuated by pretreatment with indomethacin [15, 24], it would appear that the stimulatory effect of NO is mediated by endogenous PGs. Since the HCO_3^- response to NaHS was suppressed by both indomethacin and L-NAME, it is possible that NaHS first increased NO release and then stimulated PGE_2 production, resulting in stimulation of HCO_3^- secretion.

Interestingly, the stimulatory effect of NaHS on duodenal HCO_3^- secretion was significantly mitigated by chemical ablation of capsaicin-sensitive afferent neurons. We previously reported that capsaicin, the prototypical stimulator of these neurons, increased HCO_3^- secretion in the rat duodenum and that the effect was attenuated by indomethacin, L-NAME, or sensory deafferentation [1, 25]. These results suggest that the activation of capsaicin-sensitive afferent neurons stimulates duodenal HCO_3^- secretion mediated by endogenous PGs and NO. Indeed, induction of PGE_2 production by NaHS was suppressed by both indomethacin and L-NAME, while enhancement of NO release by NaHS was prevented by L-NAME. In the present study, we did not examine the effect of sensory deafferentation on PGE_2 production and NO release in the duodenum after NaHS treatment. However, Madeirons et al. [26] reported that H_2S was gastroprotective dependent on the activation of capsaicin-sensitive afferent neurons, and capsaicin has been shown to increase PGE_2 production and NO release in the duodenum via these neurons [24, 25]; therefore, it is possible that H_2S up-regulates both PGE_2 generation and NO release in the duodenum via the activation of these afferent neurons.

H_2S has a variety of physiological effects, such as vasodilation, inhibition of leukocyte adherence, visceral analgesia, which are mediated via $K^+{}_{ATP}$ channels [3, 4, 27]. In the present study, however, the stimulatory effect of NaHS on HCO_3^- secretion and PG/NO generation in the duodenum was not affected by glibenclamide, an antagonist of $K^+{}_{ATP}$ channels. Schicho et al. [6] reported that glibenclamide was ineffective in blocking the excitatory effect of NaHS in enteric neurons. It is assumed that H_2S may have a positive effect on duodenal function via mechanisms other than the activation of $K^+{}_{ATP}$ channels.

The most important finding of the present study was that acid-induced duodenal HCO_3^- secretion was attenuated by pharmacological inhibition of CSE with propargylglycine. CSE, the enzyme responsible for the production of H_2S from L-cysteine, is mainly expressed in the liver and in vascular and nonvascular smooth muscle, but

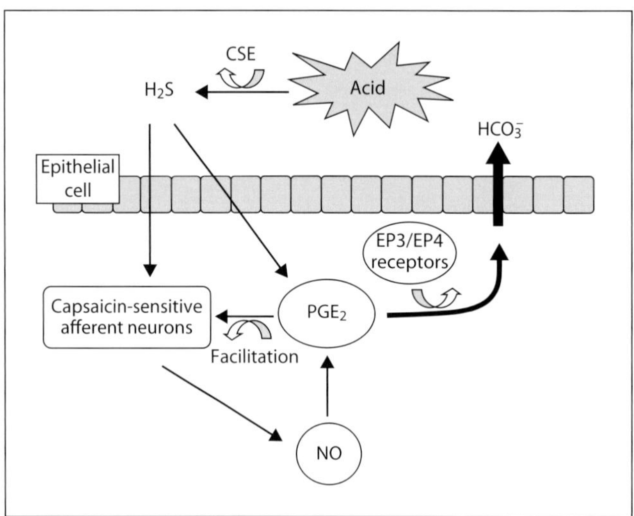

Fig. 3. Working hypothesis on the role of H_2S in the regulatory mechanism of acid-induced duodenal HCO_3^- secretion. Exogenous H_2S stimulates HCO_3^- secretion in the duodenum, and this action is mediated by endogenous PGs and NO as well as by capsaicin-sensitive afferent neurons. In addition, acidification of the mucosa somehow releases H_2S locally, probably through CSE, the enzyme responsible for the production of H_2S from L-cysteine. It is possible that H_2S stimulates these sensory neurons initially, which in turn accelerates NO release and PG production, and by so doing increases HCO_3^- secretion via the activation of EP3/EP4 receptors.

the expression of CSE protein and its enzymatic activity is also observed in the stomach and small intestine [5]. We have not confirmed the generation of H_2S in the rat duodenum following mucosal acidification, yet it is assumed that the acidification increases HCO_3^- secretion mediated, at least partly, by endogenous H_2S. It is known that acid-induced HCO_3^- secretion was inhibited by pretreatment with L-NAME and indomethacin, strongly suggesting the involvement of PGs and NO in this process [15, 28]. Indeed, many studies have demonstrated that acidification of the duodenum increases both PGE_2 and NO production together with HCO_3^- secretion [14, 16, 28, 29]. We also confirmed the increase of PGE_2 production and NO release in the duodenum after acidification and further showed that these responses were significantly blocked by propargylglycine as well as by L-NAME. Furthermore, it has been previously demonstrated that the acid-induced HCO_3^- response is also attenuated by functional ablation of capsaicin-sensitive afferent neurons and that mucosal acidification increases NO release indirectly through stimulation of these sensory neurons [13, 25, 30]. It is possible that the acidification somehow increases H_2S production, which in turn stimulates these sensory neurons, followed by upregulation of NO and PGE_2 production, and increases HCO_3^- secretion. This idea was supported by the findings that acid-induced duodenal damage was aggravated as potently as indomethacin or L-NAME, when endogenous H_2S generation was inhibited by propargylglycine.

In conclusion, H_2S stimulates HCO_3^- secretion in the duodenum, and this action is mediated by endogenous PGs and NO as well as by capsaicin-sensitive afferent neurons. It is possible that H_2S stimulates these sensory neurons initially, which in turn accelerates NO release and PGE_2 production, and by so doing increases HCO_3^- secretion (fig. 3). We also suggest that endogenous H_2S is involved in the mechanism of acid-induced HCO_3^- secretion and contributes to the maintenance of mucosal integrity in the duodenum. Further studies are needed to clarify the mechanism by which mucosal acidification increases H_2S production in the duodenum and the interactive roles of H_2S and other mediators in the acid-induced HCO_3^- secretion.

References

1 Abe K, Kimura H: The possible role of hydrogen sulfide as an endogenous neuromodulator. J Neurosci 1996;16:1066–1071.

2 Wang R: Two's company, three's a crowd: can H_2S be the third endogenous gaseous transmitter? FASEB J 2002;16:1792–1798.

3 Distrutti E, Sediari L, Mencarelli A, Renga B, Orlandi S, Antonelli E, Roviezzo F, Morelli A, Cirino G, Wallace JL: Evidence that hydrogen sulfide exerts antinociceptive effects in the gastrointestinal tract by activating K_{ATP} channels. J Pharmacol Exp Ther 2006;316:325–335.

4 Zanardo RC, Brancaleone V, Distrutti E, Fiorucci S, Cirino G, Wallace JL: Hydrogen sulfide is an endogenous modulator of leukocyte-mediated inflammation. FASEB J 2006;20:2118–2120.

5 Fiorucci S, Distrutti E, Cirino G, Wallace JL: The emerging roles of hydrogen sulfide in the gastrointestinal tract and liver. Gastroenterology 2006;131: 259–271.

6 Schicho R, Krueger D, Zeller F, Von Weyhern CW, Frieling T, Kimura H, Ishii I, De Giorgio R, Campi B, Schemann M: Hydrogen sulfide is a novel prosecretory neuromodulator in the guinea-pig and human colon. Gastroenterology 2006;131:1542–1552.

7 Kasparek MS, Linden DR, Kreis ME, Sarr MG: Gasotransmitters in the gastrointestinal tract. Surgery 2008;143:455–459.

8 Wallace JL, Vong L, McKnight W, Dicay M, Martin GR: Endogenous and exogenous hydrogen sulfide promotes resolution of colitis in rats. Gastroenterology 2009;578:569–578.

9 Linden DR, Sha L, Mazzone A, Stoltz GJ, Bernard CE, Furne JK, Levitt MD, Farrugia G, Szurszewski JH: Production of the gaseous signal molecule hydrogen sulfide in mouse tissues. J Neurochem 2008;106:1577–1585.

10 Linden DR, Levitt MD, Farrugia G, Szurszewski JH: Endogenous production of H_2S in the gastrointestinal tract: still in search of a physiological function. Antioxide Redox Signal 2010;12:1135–1146.

11 Fiorucci S, Antonelli E, Distrutti E, Rizzo G, Mencarelli A, Orlandi S, Zanardo R, Renga B, Di Sante M, Morelli A, Cirino G, Wallace JL: Inhibition of hydrogen sulfide generation contributes to gastric injury caused by non-steroidal anti-inflammatory drugs. Gastroenterology 2005;129: 1210–1224.

12 Ise F, Takasuka H, Hayashi S, Takahashi K, Koyama M, Aihara E, Takeuchi K: Stimulation of duodenal HCO_3^- secretion by hydrogen sulfide in rats: relation to prostaglandins, nitric oxide and sensory neuron. Acta Physiologica. In press.

13 Takeuchi K, Matsumoto J, Ueshima K, Okabe S: Role of capsaicin-sensitive afferent neurons in alkaline secretory response to luminal acid in the rat duodenum. Gastroenterology 1991;101:954–961.

14 Takeuchi K, Ohuchi T, Miyake H, Okabe S: Stimulation by nitric oxide synthase inhibitors of gastric and duodenal HCO_3^- secretion in rats. J Pharmacol Exp Ther 1993;266:1512–1519.

15 Sugamoto S, Kawauchi S, Furukawa O, Mimaki H, Takeuchi K: Roles of endogenous nitric oxide and prostaglandin in duodenal bicarbonate response induced by mucosal acidification in rats. Dig Dis Sci 2001;46:1208–1216.

16 Holzer P, Sametz W: Gastric mucosal protection against ulcerogenic factors in the rat mediated by capsaicin-sensitive afferent neurons. Gastroenterology 1986;91:975–987.

17 Coffey JC, Docherty NG, O'Connell PR: Hydrogen sulfide: an increasing need for scientific equipoise. Gastroenterology 2009;137:2181–2182.

18 Wallace JL: Physiological and pathophysiological roles of hydrogen sulfide in the gastrointestinal tract. Antioxid Redox Signal 2010;12:1125–1133.

19 Takeuchi K, Yagi K, Kato S, Ukawa H: Roles of prostaglandin E-receptor subtypes in gastric and duodenal bicarbonate secretion in rats. Gastroenterology 1997;113:1553–1559.
20 Aihara E, Nomura Y, Sasaki Y, Ise F, Kita K, Takeuchi K: Involvement of prostaglandin E receptor EP3 subtype in duodenal bicarbonate secretion in rats. Life Sci 2007;80:2446–2453.
21 Aoi M, Aihara E, Nakashima M, Takeuchi K: Participation of prostaglandin E receptor EP4 subtype in duodenal bicarbonate secretion in rats. Am J Physiol 2004;287:G96–G103.
22 Hu LF, Pan TT, Neo KL, Yong QC, Bian JS: Cyclooxygenase-2 mediates the delayed cardioprotection induced by hydrogen sulfide preconditioning in isolated rat cardiomyocytes. Pflügers Arch 2007;455:971–978.
23 Zhao W, Wang R: H_2S-induced vasorelaxation and underlying cellular and molecular mechanism. Am J Physiol 2002;283:H474–H480.
24 Aihara E, Kagawa S, Hayashi M, Takeuchi K: ACE inhibitor and ATI antagonist stimulate duodenal HCO_3^- secretion mediated by a common pathway: Involvement of PG, NO and bradykinin. J Physiol Pharmacol 2005;56:391–406.
25 Kagawa S, Aoi M, Kubo Y, Kotani T, Takeuchi K: Stimulation by capsaicin of duodenal HCO_3^- secretion via afferent neurons and vanilloid receptors in rats: comparison with acid-induced HCO_3^- response. Dig Dis Sci 2003;48:1850–1856.
26 Madeiros JV, Bezerra VH, Gomes AS, Barbosa AL, Lima-Junior RC, Soares PM, Brito GA, Ribeiro RA, Cunha FQ, Souza MH: Hydrogen sulfide prevents ethanol-induced gastric damage in mice: role of ATP-sensitive potassium channels and capsaicin-sensitive primary afferent neurons. J Pharmacol Exp Ther 2009;330:764–770.
27 Zhao W, Zhang J, Lu Y, Wang R: The vasorelaxant effect of H_2S as a novel endogenous gaseous K_{ATP} channel opener. EMBO J 2001;20:6008–6016.
28 Heylings JR, Garner A, Flemstrom G: Regulation of gastroduodenal HCO_3^- transport by luminal acid in the frog in vitro. Am J Physiol 1984;246:G235–G240.
29 Holm M, Johansson B, Von Bothmer C, Pettersson A, Fndriks L: Acid-induced increase in duodenal mucosal alkaline secretion in the rat involves the L-arginine/NO pathway. Acta Physiol Scand 1997;161:527–532.
30 Hogan DL, Yao B, Steinbach JH, Isenberg JI: The enteric nervous system modulates mammalian duodenal mucosal bicarbonate secretion. Gastroenterology 1993;105:410–417.

Dr. Koji Takeuchi, PhD
Division of Pathological Sciences Department of Pharmacology and Experimental Therapeutics Kyoto Pharmaceutical University Misasagi
Yamashina, Kyoto 607-8414 (Japan)
Tel. +81 075 595 4679, Fax +81 075 595 4774, E-Mail takeuchi@mb.kyoto-phu.ac.jp

Yoshikawa T, Naito Y (eds): Gas Biology Research in Clinical Practice.
Basel, Karger, 2011, pp 91–99

Hydrogen and Medical Application

Atsunori Nakao

Department of Surgery, University of Pittsburgh Medical Center, Pittsburgh, Pa., USA

Abstract

A number of basic and clinical researches have revealed that hydrogen is an important physiological regulatory factor with antioxidant, anti-inflammatory and anti-apoptotic protective effects on cells and organs, with properties to mitigate various disease status. Recently, hydrogen gas has been identified as a signaling gas molecule like nitric oxide and carbon monoxide. Therapeutic hydrogen has been applied by different delivery methods including straightforward inhalation, drinking hydrogen dissolved in water and injection with hydrogen-saturated saline. Herein, I summarize currently available data regarding the protective role of hydrogen in medicine, provide an outline of recent advances in research on the use of hydrogen as a therapeutic medical gas in diverse models of disease, and discuss the feasibility of hydrogen as a therapeutic strategy. Hydrogen's impact on therapeutic and preventive medicine could be enormous in the future.

In 2007, it was demonstrated that molecular hydrogen selectively reduces the levels of hydroxyl radicals and peroxynitrite in vitro and also exerts potent antioxidant activity in an in vivo rat cerebral ischemia model [1]. Since then, there has been accumulating evidence that hydrogen can act as a selective scavenger of hydroxyl radicals, and have potent anti-inflammatory and antiapoptotic properties [2].

Hydrogen is one very promising gaseous agent that has come to the forefront of research during the last few years. Previous studies have shown that nitric oxide (NO), carbon monoxide (CO) and hydrogen sulfide (H_2S) have cytoprotective effects in multiple experimental/clinical disease models and recognized as 'signaling gas molecules' for regulation of physiological functions. Given the recent increasing number of systems in which hydrogen is shown to mediate beneficial effects, the functions of hydrogen are likely due, in part, to the reactive oxygen species (ROS)-scavenging properties of hydrogen. However, the scavenging properties of hydrogen

Table 1. Flammability of hydrogen (H_2 content vol%)

	Lower limit	Upper limit
H_2 + air	4.6	75.5
$H_2 + O_2$	4.1	94.0

Safety standard for hydrogen and hydrogen systems, NASA 2007.

are unlikely to be the sole explanation and other, as-yet undefined mechanisms may be involved.

In this chapter, I summarize currently available data regarding the protective role of hydrogen in medicine, provide an outline of recent advances in research on the use of hydrogen as a therapeutic medical gas in diverse models of disease, and discuss the feasibility of hydrogen as a therapeutic strategy.

Chemistry of Hydrogen

Hydrogen (H_2) is the lightest molecule and a colorless, odorless, nonmetallic, tasteless, and highly combustible diatomic gas with the molecular formula H_2. Free hydrogen is comparatively rare on earth, as Earth's atmosphere contains less than 1 part per million of hydrogen. The majority of hydrogen atoms are found in water and organic compounds. Hydrogen is known to be highly reactive in the presence of specific catalysts or heat, as demonstrated in 1937 when the zeppelin Hindenburg using hydrogen for lift, ignited and was destroyed in minutes. However, safe hydrogen concentrations in air and in pure oxygen gas are <4.6 and 4.1% by volume, respectively (table 1).

Physiology of Hydrogen

Hydrogen is not endogenously produced in human cells, since enzymes with hydrogenase activity do not exist in humans. However, anaerobic organisms in the large intestine obtain their energy primarily by breaking down carbohydrates, mainly from the undigested polysaccharide fraction of plant cells and starches, via hydrogenase and generate hydrogen. Several liters of hydrogen per day are continuously produced in the human body under normal physiological conditions, primarily by the fermentation of nondigestible carbohydrates by microbiota in the large intestine [3]. Once in circulation, hydrogen is only excreted via the lungs. Therefore, exhalation of hydrogen forms the basis for a routinely used breath test for gastrointestinal transit and assessment of small intestinal bacterial overgrowth.

Protective Mechanism of Hydrogen

Although disparate mechanisms for the tissue and cellular protection afforded by hydrogen exposure have been proposed, the role of hydrogen as a scavenger of ROS has been advocated. Hydrogen selectively reduces hydroxyl radicals (•OH) and peroxynitrite ($ONOO^-$) in vitro, which are very strong oxidants that react indiscriminately with nucleic acids, lipids and proteins resulting in DNA fragmentation, lipid peroxidation, and protein inactivation. Biochemical experiments, using fluorescent probes and electron resonance spectroscopy spin traps, suggest that the effects of hydrogen against peroxynitrite are less potent than those against hydroxyl radicals [1]. The direct free radical scavenging activities of hydrogen in vivo should be more thoroughly investigated by additional studies.

Another possible mechanism underlying the cellular protection afforded by hydrogen may be an increase in antioxidant enzymes such as catalase, superoxide dismutase or heme oxygenase-1 [2, 4]. Antiapoptotic properties of hydrogen via inhibition of caspase-3 activation and induction of antiapoptotic gene B-cell lymphoma-2 (Bcl-2) have also been postulated [2]. Hydrogen exhibited anti-inflammatory activity in various injury models. Typically, oxidative stress-induced inflammatory tissue injury is inhibited by hydrogen with downregulation of proinflammatory cytokines, such as IL-1β, IL-6, chemokine (CC motif) ligand 2, and tumor necrosis factor-α [5]. Itoh et al. demonstrated using a mouse model that drinking hydrogen-rich water could attenuate an immediate-type allergic reaction by suppressing phosphorylation of FcεRI-associated Lyn and its downstream signaling molecules, which subsequently inhibited NADPH oxidase activity and reduced the generation of hydrogen peroxide [6]. This hints that the beneficial effects of hydrogen are not only imparted by its radical scavenging activity but also by modulating a specific gaseous signaling pathway.

Medical Use of Hydrogen

Experimental/Laboratory Studies

Inhalation

Inhalation of hydrogen may find clinical applications. Hydrogen inhalation therapy at a safe concentration is easily applicable for clinical settings, and may be particularly useful in the patients that are mechanically ventilated during the perioperative period. The desired concentration of hydrogen can be legitimately monitored and maintained with commercially available tools. The safety of hydrogen for humans is demonstrated by its application in Hydreliox, an exotic breathing gas mixture of 49% hydrogen, 50% helium and 1% oxygen, which is used for prevention of decompression sickness and nitrogen narcosis during very deep technical diving [7]. Hayashida et al. [8] reported that inhaled hydrogen gas at levels ranging from 0.5 to 2% limited the extent of myocardial infarction without altering hemodynamic parameters, using rat model of transient occlusion of the left anterior descending coronary artery in rats. Fukuda et al. [9] reported that inhaled hydrogen, at concentrations as low as 2–4%, reduced liver warm I/R injury in a mouse model and was associated with less oxidative stress including suppressing hepatic cell death and reducing levels of serum alanine aminotransferase and hepatic malondialdehyde. Buchholz et al. [5] have utilized an orthotopic, syngeneic, small intestinal transplant model in rats and shown that hydrogen treatment ameliorates transplant-induced intestinal injuries including mucosal erosion and mucosal barrier breakdown.

Hydrogen does not decrease the steady-state levels of nitric oxide (NO) in vitro [1]. Endogenous NO signaling pathways modulate vascular tones and leukocyte/endothelial interactions, and, therefore, it may be beneficial to spare endogenous NO. This lack of reactivity to NO may enable administration of hydrogen gas with NO, which has proven clinical effectiveness. The hyporeactivity of hydrogen with other gases at therapeutic concentrations may allow hydrogen to be administered with other therapeutic gases, including inhaled anesthesia agents. Hydrogen may be administered as a combined gas. We have recently shown that combined therapy with hydrogen and CO has enhanced therapeutic efficacy over single gas treatment for the prevention of cold ischemia/reperfusion injury after heart transplantation via both antioxidant and anti-inflammatory mechanisms [10].

Oral Intake of Hydrogen-Rich Water

Hydrogen gas by inhalation is not practical in daily life or suitable for continuous consumption for preventive or therapeutic use. In contrast, solubilized hydrogen may be beneficial since it is a portable, easily administered, and safe means of delivering molecular hydrogen [11]. Drinking hydrogen-rich water has a comparable effects to hydrogen inhalation [12]. Hydrogen water can be made by several methods, included dissolving electrolyzed hydrogen into pure water, dissolving hydrogen into water under high pressure, and by reaction of magnesium with water ($Mg + 2H_2O \rightarrow Mg(OH)_2 + H_2$). The use of magnesium hydride (MgH_2), instead of pure magnesium, may serve as more convenient storage material of hydrogen to generate hydrogen water ($MgH_2 + 2H_2O \rightarrow Mg(OH)_2 + H_2$).

Atherosclerosis and related cardiovascular diseases represent a state of inflammation and oxidative stress, characterized by the accumulation of inflammatory cells and oxidized products in affected blood vessels [13]. Ohsawa et al. [14] employed an apolipoprotein E knockout mouse model, which develops atherosclerosis in a short time due to impaired clearing of plasma lipoprotein. Oral ingestion of hydrogen-supplemented water ad libitum for 6 months prevented the development of atherosclerosis in apolipoprotein E knockout mice, in part, through its ability to limit the amount and deleterious effects of oxidative stress in the blood vessels of these mice.

Cisplatin is a potent chemotherapy agent used for treatment of a broad spectrum of malignancies, but the significant risk of nephrotoxicity frequently hinders the use of higher doses that may maximize its antineoplastic effects. Nakashima-Kamimura et al. [12] reported that hydrogen-rich water alleviates cisplatin-induced nephrotoxicity, which is mediated, in part, by the accumulation of ROS that are secondary to the ability of cisplatin to inhibit the reducing form of glutathione. Despite its protective effects against cisplatin-induced nephrotoxicity, hydrogen did not impair the antitumor activity of cisplatin against cancer cell lines in vitro or tumor-bearing mice in vivo. Therefore, hydrogen has the potential to improve the quality of life of patients during chemotherapy by efficiently mitigating the nephrotoxic side effects of cisplatin. The vast majority of late failures in renal transplantations are attributable to chronic allograft nephropathy characterized by a progressive deterioration in renal function, increasing renal hypertension and proteinuria. Cardinal et al. [15] reported that oral administration of hydrogen water prevented chronic allograft nephropathy in a rodent kidney transplantation model. Hydrogen water improved allograft function, slowed the progression of disease, reduced oxidative injury and inflammatory mediator production, and improved overall survival, doing so in part by reducing oxidative stress-induced damage and reducing the activation of mitogen-activated protein kinase signaling pathways and cytokine production [15].

Because hydrogen can easily penetrate the blood-brain barrier by gaseous diffusion, it may be a very promising agent to protect intracranial neurons. Fujita et al. [16] gave hydrogen-supplemented drinking water to the mice with Parkinson's disease induced by oral administration of 1-methyl-4-phenyl-1,2,3,6-tetrahydropy ridine (MPTP). Drinking hydrogen-containing water, even with hydrogen concentrations as low as 0.04 mM, significantly reduced the loss of dopaminergic neurons and decreased accumulation of DNA damage and lipid peroxidation, suggesting that drinking H_2-containing water may be useful in daily life to prevent or minimize the risk of lifestyle-related oxidative stress and neurodegeneration in the nigrostriatal dopaminergic pathway. Alzheimer's disease, another neurodegenerative disease, is the most common cause of progressive dementia in the elderly population and is associated with loss of cholinergic function through amyloid cascades, oxidative stress and hypoxia. The accumulation of β-amyloid in the brain, in particular accumulation of amyloid β-42, initiates a cascade of events that ultimately leads to neuronal dysfunction, neurodegeneration and dementia. Li et al. [17] reported that hydrogen-rich saline injection, intraperitoneally, daily for 2 weeks, improved the cognitive and memory functions in an amyloid β_1-42-induced Alzheimer's-like rat model, by preventing neuroinflammation and oxidative stress.

Dextran sodium sulfate (DSS)-induced rodent colitis is a well-established animal model of human inflammatory bowel disease, particularly ulcerative colitis [18]. Oral intake of hydrogen water prevented the development of DSS-induced colitis in mice [20]. The administration of H_2 remarkably reduced the clinical symptoms of DSS-induced colitis and H_2 prevented DSS-mediated destruction of epithelial crypt structure, evident by histopathological evaluation.

Injectable Hydrogen-Rich Fluid
Administration of hydrogen via an injectable hydrogen-rich vehicle may allow delivery of more accurate concentrations of hydrogen. Chen et al. [21] demonstrated that hydrogen-rich saline infusion into the tail vein significantly attenuated the severity of L-arginine-induced acute pancreatitis in rats by ameliorating increases in serum amylase activity and inhibiting neutrophil infiltration, lipid oxidation, and pancreatic tissue edema. Same group demonstrated that hydrogen treatment in rats, via tail vein injection of hydrogen-rich saline, had protective effects against intestinal contractile dysfunction and damage induced by intestinal warm I/R injury. These protective effects are possibly due to the ability of hydrogen to inhibit I/R-induced oxidative stress and apoptosis and to promote epithelial cell proliferation [22].

Eye Drops

Oharazawa et al. [23] prepared H_2-loaded eye drops (0.8 mM, pH 7.2) by dissolving hydrogen gas into saline to saturated level and then administered the H_2-loaded eye drops to the ocular surface continuously (4 liters/min) during the ischemia or reperfusion periods. The dropper, connected to the bag with H_2-loaded eye drops, was held close to the rat's eye, and drops were applied to the ocular surface. The hydrogen in the eye drops was found to immediately penetrate the vitreous body leading to elevated intravitreal hydrogen concentration after administration. Thus, H_2-loaded eye drops could effectively protect the retina from I/R injury by scavenging hydroxyl radicals and have an enormous impact via the topical application of H_2 solution.

Clinical Studies

Drinking Hydrogen Water
Metabolic syndrome is characterized by cardiometabolic risk factors that include obesity, insulin resistance, hypertension and dyslipidemia; is associated with an increased risk of developing cardiovascular disease and type II diabetes. There are several studies showing protective effects of hydrogen for metabolic disorders. Kajiyama et al. [4] treated patients with type 2 diabetes with hydrogen dissolved in water (900 ml/day) for 8 weeks. Drinking the hydrogen water resulted in reduced the levels of several biomarkers of oxidative stress, including oxidized low-density lipoprotein (LDL) cholesterol in plasma and 8-isoprostanes in urine, and improved glucose metabolism

in patients with type 2 diabetes. A pilot study conducted in Canada demonstrated that hydrogen-rich water produced by placing metallic magnesium into drinking water was also effective in attenuating oxidative stress in human subjects with potential metabolic syndrome [25]. Drinking hydrogen-rich water for an 8-week period resulted in a 39% increase in the antioxidant enzyme superoxide dismutase and a 43% decrease in thiobarbituric acid-reactive substances in urine. Further, subjects demonstrated an 8% increase in high-density lipoprotein (HDL) cholesterol and a 13% decrease in total cholesterol/HDL cholesterol from baseline to week 4. Thus, drinking hydrogen-rich water represents a potentially novel therapeutic and preventive strategy for metabolic syndrome.

Enhancing Production of Hydrogen by Intestinal Flora

Acarbose, an α-glucosidase inhibitor, is a pharmacological agent that specifically reduces postprandial hyperglycemia through retardation of disaccharide digestion, thereby reducing glucose absorption by the small intestine. Although clinical studies demonstrated that acarbose is associated with significant reduction of cardiovascular events, the mechanisms involved in prevention of cardiovascular complications were not elucidated, Suzuki et al. [26] observed 11 healthy subjects who were treated with acarbose at a dose of 300 mg/day (100 mg three times a day) and discovered that acarbose treatment significantly increased the amount of exhaled hydrogen as compared with exhaled hydrogen levels before treatment with acarbose. Based on these observations, they proposed that this α-glucosidase inhibitor may unexpectedly reduce the risk of cardiovascular disease in patients with impaired glucose tolerance or type 2 diabetes and that these benefits can be attributed, at least in part, to the ability of these drugs to neutralize oxidative stress by increasing the production of hydrogen in the gastrointestinal tract.

Hemodialysis

Nakayama et al. [27] utilized hydrogen-rich dialysis solution for hemodialysis patients and demonstrated that supplementation of hydrogen to hemodialysis solutions ameliorated inflammatory reactions. A significant decrease in systolic blood pressure (SBP) before and after dialysis was observed during the study. Significant decreases in levels of plasma monocyte chemoattractant protein 1 and myeloperoxidase were identified. During the study period, no adverse clinical signs or symptoms were observed.

Conclusions

Recent advances in experimental/clinical research for medical use of hydrogen are summarized. Due to the fact that the hydrogen studies have just begun, there is very limited information available on medical use of hydrogen and the mechanisms involved in protection, as well as functions of hydrogen in steady state physiology. However, it is undoubted that hydrogen possesses biological effects of mitigating oxidative injuries and hydrogen may have a huge potential as a safe and potent therapeutic medical gas. Further studies for detailed mechanisms or clinical human studies are certainly warranted toward clinical application.

References

1 Ohsawa I, et al: Hydrogen acts as a therapeutic antioxidant by selectively reducing cytotoxic oxygen radicals. Nat Med 2007;13:688–694.

2 Huang CS, Kawamura T, Toyoda Y, Nakao A: Recent advances in hydrogen research as a therapeutic medical gas. Free Radic Res 2010;44:971–982.

3 Levitt MD: Production and excretion of hydrogen gas in man. N Engl J Med 1969;281:122–127.

4 Kajiyama S, et al: Supplementation of hydrogen-rich water improves lipid and glucose metabolism in patients with type 2 diabetes or impaired glucose tolerance. Nutr Res 2008;28:137–143.

5 Buchholz BM, et al: Hydrogen inhalation ameliorates oxidative stress in transplantation induced intestinal graft injury. Am J Transplant 2008;8:2015–2024.

6 Itoh T, et al: Molecular hydrogen suppresses Fc-epsilon RI-mediated signal transduction and prevents degranulation of mast cells. Biochem Biophys Res Commun 2009;389:651–656.

7 Abraini JH, Gardette-Chauffour MC, Martinez E, Rostain JC, Lemaire C: Psychophysiological reactions in humans during an open sea dive to 500 m with a hydrogen-helium-oxygen mixture. J Appl Physiol 1994;76:1113–1118.

8 Hayashida K, et al: Inhalation of hydrogen gas reduces infarct size in the rat model of myocardial ischemia-reperfusion injury. Biochem Biophys Res Commun 2008;373:30–35.

9 Fukuda KI, et al: Inhalation of hydrogen gas suppresses hepatic injury caused by ischemia/reperfusion through reducing oxidative stress. Biochem Biophys Res Commun 2007;361:670–674.

10 Nakao A, et al: Amelioration of rat cardiac cold ischemia/reperfusion injury with inhaled hydrogen or carbon monoxide, or both. J Heart Lung Transplant 2010;29:544–553.

11 Cardinal JS, et al: Oral hydrogen water prevents chronic allograft nephropathy in rats. Kidney Int 2009;77:101–109.

12 Nakashima-Kamimura N, Mori T, Ohsawa I, Asoh S, Ohta S: Molecular hydrogen alleviates nephrotoxicity induced by an anti-cancer drug cisplatin without compromising anti-tumor activity in mice. Cancer Chemother Pharmacol 2009;64:753–761.

13 Stocker R, Keaney JF Jr: Role of oxidative modifications in atherosclerosis. Physiol Rev 2004;84:1381–1478.

14 Ohsawa I, et al: Consumption of hydrogen water prevents atherosclerosis in apolipoprotein E knockout mice. Biochem Biophys Res Commun 2008;377: 1195–1198.

15 Cardinal JS, et al: Oral hydrogen water prevents chronic allograft nephropathy in rats. Kidney Int 2010;77:101–109.

16 Fujita K, et al: Hydrogen in drinking water reduces dopaminergic neuronal loss in the 1-methyl-4-phenyl-1,2,3,6-tetrahydropyridine mouse model of Parkinson's Disease. PLoS One 2009;4:e7247.

17 Li J, et al: Hydrogen-rich saline improves memory function in a rat model of amyloid-beta-induced Alzheimer's disease by reduction of oxidative stress. Brain Res 2010;1328:152–161.

18 Clapper ML, Cooper HS, Chang WC: Dextran sulfate sodium-induced colitis-associated neoplasia: a promising model for the development of chemopreventive interventions. Acta Pharmacol Sin 2007;28:1450–1459.

19 Cooper HS, Murthy SN, Shah RS, Sedergran DJ: Clinicopathologic study of dextran sulfate sodium experimental murine colitis. Lab Invest 1993;69:238–249.

20 Kajiya M, Silva MJ, Sato K, Ouhara K, Kawai T: Hydrogen mediates suppression of colon inflammation induced by dextran sodium sulfate. Biochem Biophys Res Commun 2009;386:11–15.

21 Chen H, et al: Hydrogen-rich saline ameliorates the severity of L-arginine-induced acute pancreatitis in rats. Biochem Biophys Res Commun 2010.

22 Chen H, et al: The effects of hydrogen-rich saline on the contractile and structural changes of intestine induced by ischemia-reperfusion in rats. J Surg Res 2009.

23 Oharazawa H, et al: Rapid diffusion of hydrogen protects the retina: administration to the eye of hydrogen-containing saline in retinal ischemia-reperfusion injury. Invest Ophthalmol Vis Sci 2009; 51:487–492.
24 Kawasaki H, Guan J, Tamama K: Hydrogen gas treatment prolongs replicative lifespan of bone marrow multipotential stromal cells in vitro while preserving differentiation and paracrine potentials. Biochem Biophys Res Commun 2010;397:608–613.
25 Nakao A, Toyoda Y, Sharma P, Evans M, Guthrie N: Effectiveness of hydrogen rich water on antioxidant status of subjects with potential metabolic syndrome-an open label pilot study. J Clin Biochem Nutr 2010;46:140–149.
26 Suzuki Y, et al: Are the effects of alpha-glucosidase inhibitors on cardiovascular events related to elevated levels of hydrogen gas in the gastrointestinal tract? FEBS Lett 2009;583:2157–2159.
27 Nakayama M, Nakano H, Nakayama M: Novel therapeutic option for refractory heart failure in elderly patients with chronic kidney disease by incremental peritoneal dialysis. J Cardiol 2010;55:49–54.

Atsunori Nakao
Department of Surgery, University of Pittsburgh Medical Center
Pittsburgh, PA 15213 (USA)
Tel. +1 412 648 9547, Fax +1 412 624 6666, E-Mail anakao@imap.pitt.edu

Yoshikawa T, Naito Y (eds): Gas Biology Research in Clinical Practice.
Basel, Karger, 2011, pp 100–111

^{13}C-Breath Test for Studying Physiology and Pathophysiology by Using Experimental Animals

Masayuki Uchida

Food Science Institute, Division of Research and Development, Meiji Dairies Corporation, Kanagawa, Japan

Abstract

The ^{13}C-breath test has been used in various animals such as horses, piglets, dogs, rats, and mice. By using different ^{13}C-labeled compounds, it has become possible to examine the alimentary system, including *Helicobacter pylori* infection (^{13}C-labeled urea), gastric emptying ([1-^{13}C]acetic acid and [1-^{13}C]octanoic acid), gastrocecal transit time (lactose-[^{13}C]ureide), pancreatic exocrine function (N-benzoyl-L-tyrosyl-1-^{13}C-L-alanine sodium), and liver function (L-[1-^{13}C]phenylalanine). Absorption and metabolism may also be investigated by using ^{13}C-labeled amino acids. Pathophysiological conditions such as diabetes mellitus and liver cirrhosis may be investigating by using N-benzoyl-L-tyrosyl-1-^{13}C- L-alanine sodium and L-[1-^{13}C]phenylalanine, respectively. Moreover, the effects of drugs on various functions may also be estimated. The breath sample is collected using a plastic mask in large animals and a chamber system in small animals. Liquid, semisolid, or solid meals have been used as test meals. Measuring instruments include gas chromatography, isotope mass spectrometry, and infrared spectroscopy. Pharmacokinetic data are analyzed using half emptying time and lag phase. The Wagner-Nelson method is applicable in gastric emptying. The change in $^{13}CO_2$ level in the expired air, the maximum concentration, the time taken to reach the maximum concentration of $^{13}CO_2$ excretion, and the area under the curve of $^{13}CO_2$ excretion have been used for evaluation. A simple and noninvasive ^{13}C-breath-test system has been introduced in conscious rats. This method can be made applicable for the mouse. In the future, the breath test using various ^{13}C-labeled compounds in many animal species may provide us with more useful and valuable findings to clarify the physiology and pathophysiology of human beings.

In clinical settings, ^{13}C-breath tests have recently been developed as a nonradioactive alternative. The ^{13}C-labeled urea breath test, in particular, has been widely employed clinically to monitor *Helicobacter pylori* infections. In addition, the [1-^{13}C]octanoic-acid breath test has been frequently applied in the clinical diagnosis of gastric-emptying disorder since it was first reported by Ghoos et al. [1] in 1993.

Table 1. Physiological, pathophysiological and metabolic functions that can be assessed by using ^{13}C-breath tests in large animals

Animal	^{13}C-labelled compound	Purpose	Reference
Chicken	glucose	age-related glucose oxidation	Buyse et al: Life Sci 2004
Dog	octanoic acid	gastric emptying	Vegesna et al: Gastrointest Endosc 2010
Dog	octanoic acid	gastric emptying	McLellan et al: Am J Vet Res 2004
Dog	octanoic acid	gastric emptying	Yam et al: J Small Anim Pract 2004
Dog	octanoic acid	gastric emptying	Wyse et al: Res Vet Sci 2003
Dog	octanoic acid	gastric emptying	Wyse et al: Am J Vet Res 2001
Horse	acetic acid	gastric emptying	Sasaki et al: J Vet Med Sci 2005
Horse	octanoic acid	gastric emptying	Lorenzo-Figueras et al: Am J Vet Res 2005
Horse	octanoic acid	gastric emptying	Sutton et al: Equine Vet J 2003
Horse	octanoic acid	gastric emptying	Sutton et al: Equine Vet J 2002
Horse	octanoic acid	gastric emptying	Sutton et al: Equine Vet J 2002
Piglet	methacetin	liver function	Dänicke et al: Arch Anim Nutr 2008
Pony	octanoic acid	gastric emptying	Wyse et al: Equine Vet J 2001
Tiger and Cheetah	urea	*H. pylori* infection	Chatfield et al: J Zoo Wildl Med 2004

Gastric-emptying time can be indirectly measured by monitoring the $^{13}CO_2$ concentration in expired air after a ^{13}C-labeled substance such as [1-^{13}C]acetic or [1-^{13}C] octanoic acid has been ingested and absorbed from the small intestine.

In experimental settings, breath tests have been used to investigate various functions such as gastric emptying, liver function, pancreatic exocrine function, and numerous pathological conditions, including gastritis, hepatectomy, liver cirrhosis, and streptozotocine-induced diabetes, in various animals such as horses, dogs, piglets, tigers and cheetahs (table 1). Thus, the application of ^{13}C-labeled substances for investigating various physiological functions and pathological states was introduced.

Rats and mice are generally used for experimental procedures, and many assessments of the breath test in small animals such as rats and mice have been reported (table 2). However, a simple and noninvasive breath test in such small animals has not been clearly established. Therefore, the present author introduced a simple and noninvasive system for monitoring expired $^{13}CO_2$ in conscious rats using a desiccator and a pump, both of which are readily available experimental instruments, and an infrared spectrometer (UBiT-IR300; Otsuka Electronics, Co. Ltd., Osaka, Japan), which is

Table 2. Physiological, pathophysiological and metabolic functions that can be assessed by using ^{13}C-breath tests in small animals

Animal	^{13}C-labelled compound	Purpose	Reference
Mouse	acetic acid	gastric emptying	Matsumoto et al: Biol Pharm Bull 2008
Mouse	amino acid	endotoxemia	Butz et al: Rapid Commun Mass Spectrom 2009
Mouse	octanoic acid	gastric emptying	Symonds et al: Clin Exp Pharmacol Physiol 2000
Mouse	urea	*H. pylori* infection	Hammond et al: Helicobacter 1999
Rat	acetic acid	gastric emptying	Uchida et al: J Pharmacol Sci 2005
Rat	acetic acid octanoic acid	gastric emptying	Uchida and Shimizu: Biol Pharm Bull 2007
Rat	galactose	liver function in STZ rat	Mion et al: Can J Physiol Pharmacol 1999
Rat	lactose ureide	gastrocecal transit time	Uchida and Yoshida: J Pharmacol Sci 2009
Rat	leucine	postprandial phase	Bujko et al: Br J Nutr 2007
Rat	methacetin	hepatic injury	Shirin et al: J Gastroenterol Hepatol 2008
Rat	methionine	liver regeneration	Chatfield et al: J Zoo Wildl Med 2004
Rat	N-benzoyl-L-tyrosyl-alanine	pancreatic function	Uchida and Mogami, Biol Pharm Bull 2008
Rat	N-benzoyl-L-tyrosyl-alanine	pancreatic function	Kohno et al: Scand J Gastroenterol 2007
Rat	octanoic acid	gastric emptying	Schoonjans et al: Neurogastroenterol Motil 2002
Rat	octanoic acid	liver chirrosis	Shalev et al: Dig Dis Sci 2010
Rat	orinitine	hepatectomy	McLellan et al: Am J Vet Res 2004
Rat	phenylalanine	CCl4-induced cirrhosis	Yan et al: Life Sci 2006
Rat	phenylalanine	CCl4-induced cirrhosis	Yan et al: Eur J Clin Invest 2005
Rat	phenylalanine	hepatectomy	Ishii et al: Surgery 2002
Rat	phenylalanine	liver function	Yan et al: Zhonghua Gan Zang Bing Za Zhi 2005
Rat	phenylalanine	phenyalalnine hydroxylase	Ito et al: Digestion 2001
Rat	phenylalanine methionine, etc.	hepatectomy	Ishii et al: Nippon Yakurigaku Zasshi 2002
Rat	sucrose	mucositis	Tooley et al: Cancer Chemother Pharmacol 2010
Rat	sucrose	mucositis	Clarke et al: Cancer Biol Ther 2006
Rat	triglyceride	lipid metabolism	Michalski et al: Eur J Nutr 2005
Rat	urea	disposition of urea	Nomura et al: Arzneimittelforsch 2006

marketed for the diagnosis of *H. pylori* infections and can be operated without any technical expertise [2].

Breath Test in Experimental Animals

Various animals have been used for the breath test thus far, but most research is conducted using rats and mice. The horse, pony, Sumatran tiger, cheetah, piglet, dog, and broiler chicken have also been used in veterinary science (table 1). Vegesna et al. [3] reported the efficacy of endoscopic pyloric suturing to facilitate weight loss in dogs by investigating gastric emptying using [1-^{13}C]octanoic acid breath tests. Buyse et al. [4] observed the change in age-related glucose oxidation rates in broiler chickens by using ^{13}C-labeled glucose. In the case of insects, Voigt et al. [5] reported that female bush crickets fuel their metabolism with male nuptial gifts, supporting the theory that females quickly use these nutrients for metabolism, thus receiving immediate benefits from spermatophore feeding.

Test Meals for the Breath Test in Animals

In clinical practices, [1-^{13}C]acetic acid has been used for the evaluation of gastric emptying in the form of a liquid test meal such as coffee or juice. In contrast, [1-^{13}C] octanoic acid has been used in the form of a solid test meal such as egg yolk or steak for the evaluation of gastric emptying in humans. Solid or semisolid test meals may be used in large animals such as the horse, pony, or dog. In small experimental animals, a solid meal has been supposed to be difficult to administer as a test meal. However, Symonds et al. [6] described the use of a solid meal consisting of an egg yolk and solid mouse chow containing [1-^{13}C]octanoic acid, all of which was consumed within 1 min. This consumption time may affect the variation in the results of the breath test.

Symonds et al. [6] used Intralipid (Fresenius Kabi, Sweden), whereas the present author [2] used Racol (Ohtsuka Pharmaceutical Co. Ltd., Japan) as the liquid test meal. In clinical practice, enteral nutritional formulations have been used as liquid test meals. Therefore, these enteral nutritional formulations are also useful for liquid test meals in animal experiments.

Collection of Expired Breath Samples

In large animals such as horses, piglets or dogs, a plastic mask connected to a sampling bag is generally used for collecting expired air samples. Schoonjans et al. [7] reported the use of the breath test with [1-^{13}C]ocatanoic acid in rats placed in metabolic airtight cages. This method is stressful for the rats. In their method, the breath tests were performed by a fully automated system of computer-guided switching valves within metabolic cages [7]. These instruments are expensive. However, in small animals such as cats, guinea pigs, rats, and mice, the chamber system is useful for collecting expired breath without stress and anesthesia. The present author used a desiccator as an animal chamber, and expired air was collected by a tube and aspiration pump, which were fairly inexpensive [2].

Measuring Expired Breath Samples

In general, breath samples are analyzed using gas chromatography and isotope mass spectrometry; the latter is the established standard method for the sensitive and accurate measurement of the $^{13}CO_2/^{12}CO_2$ ratio in breath samples [8]. These analytical apparatuses are expensive, and expert knowledge is required to operate the apparatuses. Ohara et al. [9] and Leodolter et al. [10] have developed a new compact instrument for nondispersive isotope-selective infrared spectroscopy (UBiT-IR300). By using this apparatus, the $^{13}CO_2$ concentration in the expired air can be measured simply by placing the breath-sampling bags in the sample joint, without the need for technical procedures. Kato et al. [11] reported that the data obtained from mass spectrometry were significantly correlated with the data from infrared spectroscopy ($r = 0.99989$; $p < 0.0001$).

Pharmacokinetic Data Analysis

Ghoose et al. [1] suggested the half emptying time and lag phase as evaluation parameters in the gastric emptying test using [1-^{13}C]octanoic acid in humans. Sanaka et al. [12] reported the usefulness of the Wagner-Nelson method for the analysis of gastric emptying.

The changes in expired $^{13}CO_2$ level over time ($\Delta^{13}CO_2$) have also been shown to be useful analytical data [2]. The maximum concentration (C_{max}; ‰), the time taken to reach the maximum concentration (T_{max}; min), and the area under the curve (AUC; ‰·min) calculated using the $\Delta^{13}CO_2$ values, all of which are usually used as pharmacokinetic parameters of drugs, have also been used as evaluation parameters in the breath test.

Breath Test for Evaluating *H. pylori* Infection

The urea breath test has been used widely for the diagnosis of *H. pylori* infection in humans. The same methods have also been used in rats and mice for the diagnosis of *H. pylori* infection. Chatfield et al. [13] validated the ^{13}C-urea breath test using

Sumatran tigers *(Panthera tigris)* and cheetahs *(Acinonyx jubatus)*, and established the technique as a valuable, simple, accurate, and sensitive tool for monitoring the eradication of *H. pylori* during therapy for gastritis. Using the ^{13}C-urea breath test, Plonka et al. [14] reported that *H. pylori* infection in Polish shepherds originates from sheep.

Breath Test for Evaluating Gastric Emptying

[1-^{13}C]Acetic acid and [1-^{13}C]octanoic acid have been used for the evaluation of gastric emptying of liquid test meals and solid test meals, respectively. Various animals such as horses, dogs, rats and mice have been used for evaluating gastric emptying (tables 1, 2).

When comparing [1-^{13}C]acetic acid and [1-^{13}C]octanoic acid in gastric emptying, the present author [15] reported that [1-^{13}C]acetic acid changed to a significantly greater extent than [1-^{13}C]octanoic acid when administered at the same molar dosage in a liquid test meal.

Breath Test for Evaluating Liver Function

Some ^{13}C-labelled compounds have been used for the analysis of liver function. Yan et al. [16] reported a quantitative liver function assessment using L-[1-^{13}C]phenylalanine in rats. Ishii et al. [17] evaluated liver regeneration using L-[1-^{13}C]methionine in a rat model of 70% hepatectomy. The effects of hepatectomy have also been tested in rats using ^{13}C-ornitine, L-[1-^{13}C]phenylalanine, and L-[1-^{13}C]methionine (table 2). Dänicke et al. [18] reported the effects of a wet preservation of triticale contaminated mainly with deoxynivalenol with sodium metabisulphite on the liver function of piglets by using ^{13}C-methacetine. L-[1-^{13}C]Phenylalanine and [1-^{13}C]octanoic acid have also been used in a carbon tetrachloride-induced cirrhosis model.

Breath Test for Evaluating Pancreatic Exocrine Function

In 2007, Kohno et al. [19] synthesized ^{13}C-dipeptide, N-benzoyl-L-tyrosyl-1-^{13}C-L-alanine sodium, and evaluated pancreatic exocrine function by administering N-benzoyl-L-tyrosyl-1-^{13}C-L-alanine sodium dissolved in distilled water to anesthetized or conscious rats. It is assumed that orally administered N-benzoyl-L-tyrosyl-1-^{13}C-L-alanine sodium is evacuated to the duodenum, where it is cleaved by one of the pancreatic proteases, carboxypeptidase, thereby releasing L-[1-^{13}C]alanine, which is absorbed from the intestine and metabolized in the liver to $^{13}CO_2$. The present author [20] also used N-benzoyl-L-tyrosyl-1-^{13}C-L-alanine sodium for the evaluation of pancreatic exocrine secretion and diagnosis of alloxan-induced diabetes mellitus.

Breath Test for Evaluating Sugar and Lipid Metabolism

[1-^{13}C]Glucose and [1-^{13}C]fructose as monosaccharides, and [1-$^{13}C^{glucose}$]sucrose, [1-$^{13}C^{fructose}$]sucrose, [1-$^{13}C^{glucose}$]paratinose and [1-$^{13}C^{fructose}$]paratinose as disaccharides have been used for the evaluation of sugar metabolism in rats, based on the theory that paratinose metabolizes more gradually compared with sucrose [21]. Buyse et al. [4] reported glucose oxidation using ^{13}C-labeled glucose in broiler chickens.

Michalski et al. [22] assessed lipid metabolism using ^{13}C-triglyceride in rats.

Breath Test for Evaluating Gastrocecal Transit Time

In 1995, Heine et al. [23] reported the assessment of orocecal transit time in humans using lactose-[^{13}C]ureide in humans. After passing the small bowel, the sugar-urea bond of lactose-[^{13}C]ureide is split by the allantoicase of bacteria colonizing the cecum. Split urea is absorbed from the colon and metabolized in the liver to $^{13}CO_2$. Ruemmele et al. [24] reported that the human gut tissue possesses no allantoicase-like activity. In experimental animals, only 1 study has been reported in which the present author [25] found that a priming dose of lactose-ureide was needed to investigate gastrocecal transit time in rats. In a clinical study, Wutzke and Schütt [26] also reported that a priming dose of lactose-ureide was necessary to activate the flora metabolizing lactose-[^{13}C]ureide, for the investigation of orocecal transit time.

Breath Test for Evaluating Cachexia

In the only study concerning cachexia, Butz et al. [27] reported the effects of endotoxemia using various amino acids in mice and found that the exhaled breath $\Delta^{13}CO_2$ value was a valuable real-time biomarker of cachexia associated with an acute phase response due to endotoxemia.

Simple and Noninvasive Breath Test in Rats

A simple and noninvasive breath test was performed in male Sprague-Dawley rats weighing 200–250 g that were fasted in mesh cages for 18 h before each experiment

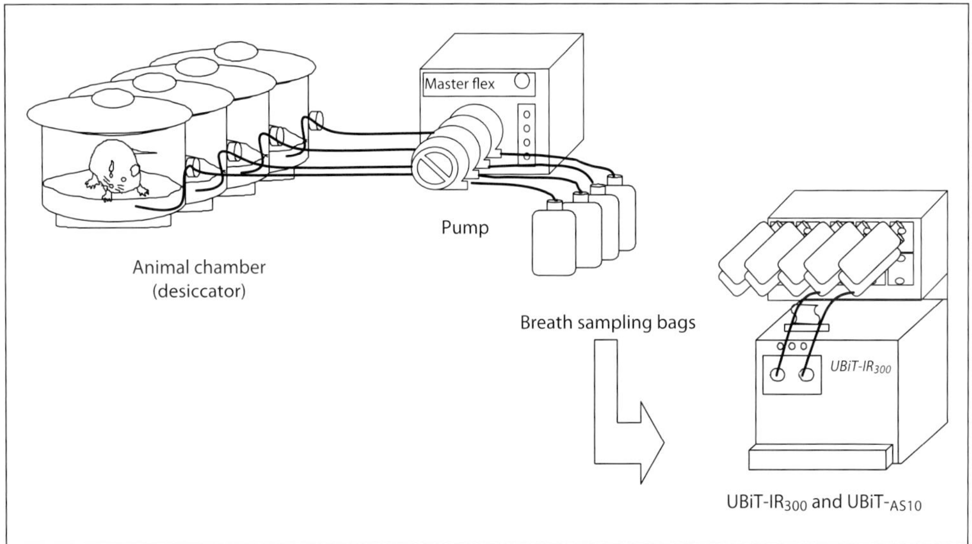

Fig. 1. Schematic illustration of the system used for monitoring $^{13}CO_2$ levels in the expired air from rats. Briefly, this system comprised a desiccator that was used as an animal chamber, a pump, and a breath-sampling bag. $^{13}CO_2$ level was measured with an infrared spectrometer (UBiT-IR300) and an automatic sampler (UBiT-AS10). Aspirating the expired air caused fresh air to automatically flow into the desiccator to replace expired air through a hole in the side of the chamber [2].

in order to prevent coprophagy, but were allowed free access to drinking water during this period [2].

A schematic illustration of the system used for monitoring $^{13}CO_2$ levels in expired air from rats is shown in figure 1. The desiccator and aspiration pump were selected because they were easy to set up and relatively inexpensive. The UBiT-IR300 and UBiT-$_{AS10}$ apparatus were chosen as they allowed $^{13}CO_2$ to be measured simply and effectively. A desiccator with a volume of 2,000 ml was employed, so that the rats could move freely within the chamber, and the expired air could be collected effectively in the breath-sampling bag. Aspirating the expired air caused fresh air to be automatically drawn into the desiccators through a hole, through which the aspiration tube also passes, on the side of the desiccators (fig. 1). The air in the chamber was continuously aspirated during the experimental period. Aspirated air was discharged outside this breath-test system, except during the collection of expired air in the breath-sampling bag.

Racol containing the ^{13}C-labeled compound was used as the test meal, to ensure consistency with previous clinical studies.

The aspiration volume was determined based on the fact that a volume of 75 ml/min has previously been used for the artificial respiration of anesthetized rats weighing 200–250 g. Similar respiratory patterns were seen at all of the aspiration volumes tested. However, the rats demonstrated slight anoxia at a volume of 75 ml/min. Anoxia was not observed at volumes of 150 ml/min or greater. An aspiration volume of 150 ml/min was therefore used in our study.

The rats were placed in the chamber immediately after oral administration of the test meal, and expired air was collected at appropriate intervals. At each sampling point, the expired air was collected into a breath-sampling bag for 1.5 min when the aspiration volume was 150 ml/min. This system allowed 4 rats to be tested simultaneously under the same conditions (fig. 1). $^{13}CO_2$ levels in the expired air were measured by placing the breath-sampling bags into the sample joint of the UBiT-IR300 infrared analyzer (fig. 1). The measured values were presented as the $\Delta^{13}CO_2$ (‰).

Gastric emptying was evaluated by using [1-^{13}C]acetic acid. The optimal dose of [1-^{13}C]acetic acid was 16 mg/kg and the optimum volume of the test meal was 2.5–5 ml/kg. Metoclopramide has been shown to enhance gastrointestinal motility in both animals and humans. In this study, metoclopramide dose-dependently and significantly enhanced gastric emptying. Metoclopramide significantly increased the C_{max} value at a dose of 3 mg/kg and significantly shortened the T_{max} values at doses of 1 and 3 mg/kg [2].

In contrast, atropine sulfate has been shown to inhibit gastrointestinal motility. Atropine sulfate dissolved in distilled water was administered subcutaneously (0.03–0.3 mg/kg; 1 ml/kg) 30 min before the administration of the test meal. The C_{max} values were significantly diminished in atropine sulfate-treated rats, but leveled off at 0.1 and 0.3 mg/kg doses. The T_{max} value was dose-dependently delayed in atropine sulfate-treated rats, but also leveled off at the 0.1- and 0.3-mg/kg doses. Atropine sulfate did not significantly affect AUC_{120min} values at either dose used in this study [28].

Gastriocecal transit time was also tested by using this breath test system [25], but a priming dose of lactose-ureide was needed. Three doses of lactose-[^{13}C]ureide (30, 60, and 120 mg/kg) were administered orally and the breath test was performed. Gastrocecal transit time at doses of 30, 60 and 120 mg/kg were 160 ± 24.5, 150 ± 17.3 and 180 ± 0.0 min, respectively, showing no significant difference between the 3 doses (fig. 2). To determine the effect of atropine sulfate on gastrocecal transit time, 2 doses of atropine sulfate (0.1 and 1 mg/kg; 1 ml/kg) were administered subcutaneously 30 min before the breath test [28]. At a dose of 1 mg/kg, a significant delay in gastrocecal transit time was observed. Moreover, significantly lower $^{13}CO_2$ levels in expired air were observed.

Recently, the present author reported a method of simultaneous measurement of gastric emptying and gastrocecal transit time by using [1-^{13}C]acetic acid and lactose-[^{13}C]ureide, respectively, in rats [29].

Application of the Simple and Noninvasive Breath Test for Mice

This simple and noninvasive breath test system is also applicable for mice [30]. For mice, the volume of the animal chamber and the aspiration volume of the pump were

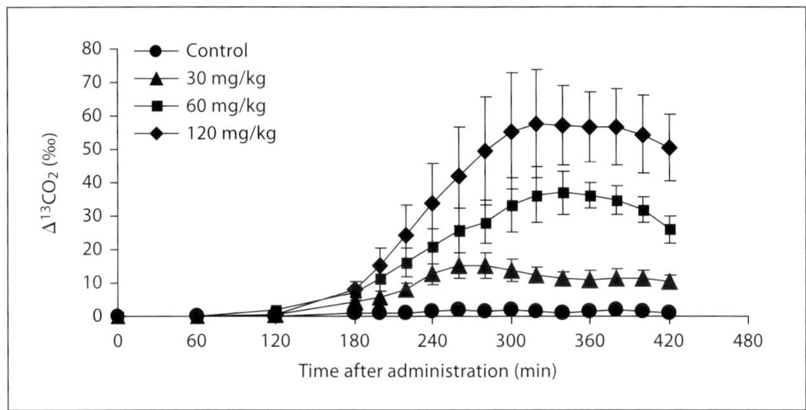

Fig. 2. Effect of the dose of N-benzoyl-L-tyrosyl-1-^{13}C-L-alanine sodium on the change in $^{13}CO_2$ level in expired air from rats. Values represent the mean ± SEM (n = 4) [25].

changed to 500 and 50 ml/min, respectively. The effects of loperamide, morphine, mosapride, and itopride on gastric emptying have been investigated [30]. This method may be applicable for animals larger than the rat, such as guinea pigs and rabbits, by using larger desiccators and increasing the aspiration volume of the pump.

Conclusion

In summary, the noninvasive and simple breath test using ^{13}C-labeled compounds is useful for the evaluation of physiological functions and pathological states. In future, the synthesis of various new ^{13}C-labeled compounds and the use of various experimental animals may provide us with more useful and valuable clues concerning the physiology and pathophysiology of humans.

References

1 Ghoos YF, Maes BD, Geypens BJ, Mys G, Hiele MI, Rutgeerts PJ, Vantrappen G: Measurement of gastric emptying rate of solids by means of a carbon-labeled octanoic acid breath test. Gastroeneterology 1993;104:1640–1647.

2 Uchida M, Endo N, Shimizu K: Simple and noninvasive breath test using ^{13}C-acetic acid to evaluate gastric emptying in conscious rats and its validation by metoclopramide. J Pharmacol Sci 2005;98:388–395.

3 Vegesna A, Korimilli A, Besetty R, Bright L, Milton A, Agelan A, McIntyre K, Malik A, Miller L: Endoscopic pyloric suturing to facilitate weight loss: a canine model. Gastrointest Endosc 2010;72:427–431.

4 Buyse J, Geypens B, Malheiros RD, Moraes VM, Swennen Q, Decuypere E: Assessment of age-related glucose oxidation rates of broiler chickens by using stable isotopes. Life Sci 2004;75:2245–2255.

5 Voigt CC, Kretzschmar AS, Speakman JR, Lehmann GU: Female bush crickets fuel their metabolism with male nuptial gifts. Biol Lett 2008;4:476–478.
6 Symonds E, Butler R, Omari T: The effect of the GABAB receptor agonist baclofen on liquid and solid gastric emptying in mice. Eur J Pharmacol 2003;470:95–97.
7 Schoonjans R, Van Vlem B, Van Heddeghem N, Vandamme W, Vanholder R, Lameire N, Lefebvre R, De Vos M: The ^{13}C-octanoic acid breath test: validation of a new noninvasive method of measuring gastric emptying in rats. Neurogastroenterol Motil 2002;14:287–293.
8 Craig H: Isotope standards for carbon and oxygen and correlation factors for mass spectrometric analysis of carbon dioxide. Geochim Cosmichim Acta 1957;12:133–149.
9 Ohara S, Kato M, Asaka M, Toyama T: The U-BiT-100 $^{13}CO_2$ infrared analyzer: comparison between infrared spectrometric analysis and mass spectrometric analysis. Helicobacter 1998;3:49–53.
10 Leodolter A, Arnim UV, Gerards C, Kahl S, Malfertheiner P: Pre- and post-treatment diagnosis of *H. pylori* infection by the ^{13}C-urea breath test (^{13}C-UBT): variation of a new isotope-selective infrared spectrometer (ISIS) (abstract). Gastroenterology 2000;118(suppl 2):A506.
11 Kato M, Saito M, Fukuda S, Kato C, Ohara S, Hamada S, Nagashima R, Obara K, Suzuki M, Honda H, Asaka M, Toyota T: ^{13}C-Urea breath test, using a new compact nondispersive isotope-selective infrared spectrophotometer: comparison with mass spectrometry. J Gastroeneterol 2004;39:629–634.
12 Sanaka M, Yamamoto T, Ishii T, Kuyama Y: The Wagner-Nelson method can generate an accurate gastric emptying flow curve from CO_2 data obtained by a ^{13}C-labeled substrate breath test. Digestion 2004;69:71–78.
13 Chatfield J, Citino S, Munson L, Konopka S: Validation of the ^{13}C-urea breath test for use in cheetahs (*Acinonyx jubatus*) with *Helicobacter*. J Zoo Wildl Med 2004;35:137–141.
14 Plonka M, Bielanski W, Konturek SJ, Targosz A, Sliwowski Z, Dobrzanska M, Kaminska A, Sito E, Konturek PC, Brzozowski T: *Helicobacter pylori* infection and serum gastrin, ghrelin and leptin in children of Polish shepherds. Dig Liver Dis 2006;38:91–97.
15 Uchida M, Shimizu K: ^{13}C-acetic acid is more sensitive than ^{13}C-octanoic acid for evaluating gastric emptying of liquid enteral nutrient formula by breath test in conscious rats. Biol Pharm Bull 2007;30:487–489.
16 Yan W, Xiong P, Liu Z, Huang G: Results of L-[1-^{13}C]phenylalanine breath test with air isotope ratio mass spectrometry can reflect the activity of phenylalanine hydroxylase in cirrhotic rat liver. Rapid Commun Mass Spectrom 2006;20:602–608.
17 Ishii Y, Asai S, Kohno T, Ito A, Iwai S, Ishikawa K: Recovery of liver function in two-third partial hepatectomized rats evaluated by L-[1-^{13}C]phenylalanine breath test. Surgery 2002;132:849–856.
18 Dänicke S, Beineke A, Goyarts T, Valenta H, Beyer M, Humpf HU: Effects of a Fusarium toxin-contaminated triticale, either untreated or treated with sodium metabisulphite (Na_2S_2O5, SBS), on weaned piglets with a special focus on liver function as determined by the ^{13}C-methacetin breath test. Arch Anim Nutr 2008;62:263–286.
19 Kohno T, Ito A, Hosoi I, Hirayama J, Shibata K: Synthetic ^{13}C-dipeptide breath test for the rapid assessment of pancreatic exocrine insufficiency in rats. Scand J Gastroenterol 2007;42:992–999.
20 Uchida M, Mogami O: Usefulness of breath test for evaluating pancreatic exocrine function using N-benzoyl-L-tyrosyl-1-^{13}C-L-alanine sodium in non-invasive and conscious rats. Biol Pharm Bull 2008;31:785–788.
21 Tonouchi T, Yamaji T, Uchida M, Koganei M, Sasayama A, Kaneko T, Urita, Okuno M, Suzuki K, Kashimura J, Sasaki H: Studies on absorption and metabolism of palatinose (isomaltulose) in rats. Br J Nutr 2010, in press.
22 Michalski MC, Briard V, Desage M, Geloen A: The dispersion state of milk fat influences triglyceride metabolism in the rat: a $^{13}CO_2$ breath test study. Eur J Nutr 2005;44:436–444.
23 Heine WE, Berthold HK, Klein PD: A novel stable isotope breath test: ^{13}C-labeled glycosyl ureides used as noninvasive markers of intestinal transit time. Am J Gastroenterol 1995;90:93–98.
24 Ruemmele FM, Heine WE, Keller KM, Lentze MJ: Metabolism of glycosyl ureides by human intestinal brush border enzymes. Biochim Biophys Acta 1997;1336:275–280.
25 Uchida M, Yoshida K: Non-invasive method for evaluation of gastrocecal transit time by using a breath test in conscious rats. J Pharmacol Sci 2009;110:227–230.
26 Wutzke KD, Schütt M: The duration of enzyme induction in orocaecal transit time measurements. Eur J Clin Nutr 2007;61:1162–1166.
27 Butz DE, Cook ME, Eghbalnia HR, Assadi-Porter F, Porter WP: Changes in the natural abundance of $^{13}CO_2/^{13}CO_2$ in breath due to lipopolysacchride-induced acute phase response. Rapid Commun Mass Spectrom 2009;23:3729–3735.

28 Uchida M, Yoshida K, Shimizu K: Effect of atropine sulfate on gastric emptying and gastrocecal transit time evaluated by using the [1-^{13}C]acetic acid and actose-[^{13}C]ureide breath test in conscious rats. J Breath Res 2009;3:1–4.
29 Uchida M, Yoshida K: Simultaneous measurement of gastric emptying and gastrocecal transit time by using rat breath test. J Pharmacol Sci 2010;112 (suppl 1):252P.
30 Matsumoto K, Kimura H, Tashima K, Uchida M, Horie S: Validation of ^{13}C-acetic acid breath test by measuring effects of loperamide, morphine, mosapride, and itopride on gastric emptying in mice. Biol Pharm Bull 2008;31:1917–1922.

Masayuki Uchida, PhD
Food Science Institute, Division of Research and Development, Meiji Dairies Corporation
540 Naruda, Odawara
Kanagawa 250-0862 (Japan)
Tel. +81 465 37 3664, Fax +81 465 36 2776, E-Mail masayuki_uchida@meiji-milk.com

Yoshikawa T, Naito Y (eds): Gas Biology Research in Clinical Practice.
Basel, Karger, 2011, pp 112–118

Carbon-13 and Its Clinical Application

Yusuke Tando[a] · Atsufumi Matsumoto[a] · Yuki Matsuhashi[a] · Hikaru Tanaka[a] · Miyuki Yanagimachi[a] · Teruo Nakamura[b]

[a]Department of Endocrinology and Metabolism, Hirosaki University School of Medicine, and [b]Hirosaki University School of Health Science, Aomori, Japan

Abstract

The stable isotope carbon-13 (^{13}C) is widely used in physiological investigation and medical research. Compared with a radioactive isotope tracing (e.g. ^{14}C and ^{3}H), it has neither radiation damages nor radiolysis effects. Dynamic processes of substrate metabolism can be estimated by calculating enrichment of the particular substrate labeling with ^{13}C or by measuring $^{13}CO_2$:$^{12}CO_2$ isotope ratio in breathing. Among the many applications of ^{13}C-labeled tracer technique, the ^{13}C-urea breath test is the most conspicuous diagnostic method for clinical practice. Other substrates labeling with ^{13}C are also administered orally or intravenously, but they are still relatively unknown as noninvasive diagnostic tools in clinical practice. In order to promote these new simple techniques, we need to compare them with traditional complicated techniques and to establish the standard procedure. This review aims to introduce and summarize the clinical applications of ^{13}C with typical examples, especially focusing on the current use of ^{13}C in the gas biology.

Since early times, a minimally invasive procedure has been generally required in medical practice. In recent years, many new less-invasive techniques, such as Doppler ultrasonography and magnetic resonance imaging, have been applied to medicine, providing both physiological and functional information that may be available for clinical diagnosis and treatment. A stable isotope tracer technique, as one of the less-invasive techniques, has also been employed since the 1970s in the study of glucose metabolism [1], pancreatic function [2], liver function [3] and bacterial overgrowth [4]. With recent advances in both tracer chemistry and mass spectrometer technology, this technique has become more easily available. Carbon-13 (^{13}C) is the most representative stable isotope from a practical point of view. Many types of substrates labeling with ^{13}C are administered orally or intravenously without health hazards of radioactivity [5] (table 1). In this chapter, clinical applications of ^{13}C-labeled substrate, especially for diagnostic purposes, are reviewed.

Table 1. Substrates of ^{13}C breath tests

Substrate	Purpose of examination
Gastric and duodenal disease	
Octanoic acid	gastric emptying (solids)
Acetate	gastric emptying (liquids)
Urea	*H. pylori* infection
Pancreatic disease	
Trioctanoin	fat malabsorption
Triolein	fat malabsorption
Hiolein	fat malabsorption
Mixed triglyceride	fat malabsorption
Liver disease	
Aminopyrine	activity of microsomal monoxygenases
Caffeine	hepatic microsomal biotransformation
Galactose	liver fibrosis (HBV), Galactosemia
Glucose	glucose metabolism, diabetes, obese, insulin resistance, glycogen metabolism
Phenylalanine	hepatocyte function, cytosolic enzyme activity
Leucine	hepatocyte function (protein turnover)
Glutamate	gluconeogenesis
Intestinal disease	
Lactose	lactase deficiency
Xylose	small intestine bacterial overgrowth
Bile acid	small intestine bacterial overgrowth
Ureide (-cellobiose)	orocecal transit time
Others	
Medium- and long-chain triglyceride	primary carnitine deficiency
(Alpha)-linoleic acid	fatty acid metabolism
Uracil	activity of dihydropyrimidine dehydrogenase

^{13}C in Medical Practice – To the Present

The ^{13}C-labeled tracer technique has developed after clinical use of a ^{14}C-labeled 'radioactive' tracer technique [6]. The 'nonradioactivity' is a substantial advantage of ^{13}C-labeled substrate over any other radioactive tracer, which can be valuable in certain clinical situations, such as in infants, children, and women of childbearing age. ^{13}C is a naturally occurring isotope present to the extent of approximately 1.1% of the major isotopic species, ^{12}C [7]. The ^{13}C-labeled tracer technique is based on that very fact, though this ratio differs slightly according to the place of residence or dietary habits [7]. The ^{13}C-labeled tracer technique is divided into two major methods. One is dilution analysis and the other is the breath test. The dilution technique is

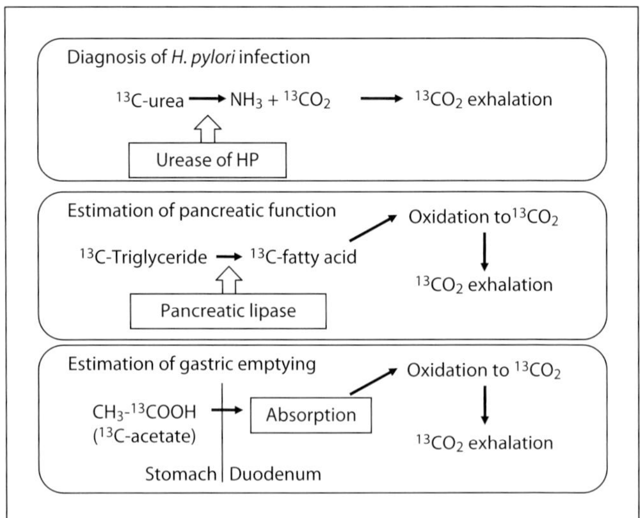

Fig. 1. Schematic depiction of the ^{13}C breath tests.

employed for kinetic assessment in the field of pharmacology and nutritional science [8]. On the other hand, breath tests have been employed in the study of gastric emptying, liver function, fat absorption, and so on. The methodology and the condition of the ^{13}C breath test were almost established by Schoeller and Watkins [9] in 1970s. The breath tests are usually carried out as follows: the subject takes in a certain amount of carbon compound containing carbon labeled with its stable isotope ^{13}C and the enrichment of the particular substrate labeling with ^{13}C or the $^{13}CO_2$:$^{12}CO_2$ isotope ratio is measured in the expired air (fig. 1).

Helicobacter pylori *Infection*

H. pylori is a Gram-negative pathogen that colonizes the stomach. *H. pylori* infection induces chronic gastritis and is known risk factor for peptic ulcer and gastric adenocarcinoma, and non-Hodgkin lymphoma [10]. The ^{13}C-urea breath test (^{13}C-UBT) is a noninvasive method for detecting *H. pylori* infection, which is possible to perform without endoscopy. The ^{13}C-UBT exploits the enzyme urease produced by *H. pylori*, as this enzyme hydrolyses the orally administered ^{13}C-labeled urea into ammonia and labeled CO_2, which is absorbed and transported to lung for excretion (fig. 1). The sensitivity and specificity are very high and range from 90 to 98% and from 92 to 100% [11, 12]. Comparing other tests for detecting *H. pylori* infection such as biopsy-based tests, the advantage of the ^{13}C-UBT is that it samples the whole stomach even within the gastric mucosa. The adverse effect of proton pump inhibitors, high dose of H_2 blockers, and various conditions for testing on ^{13}C-UBT have been reported [12].

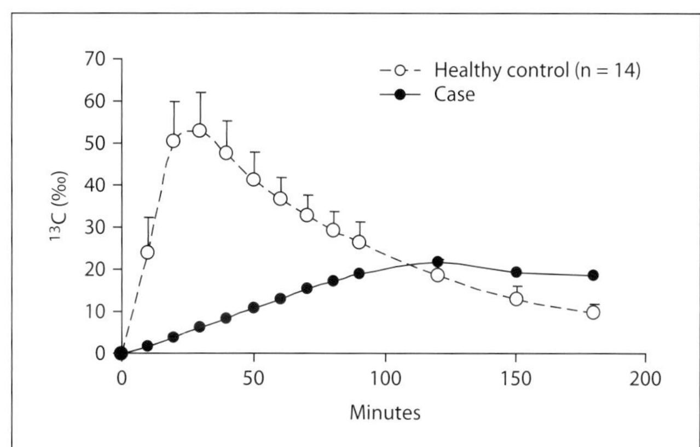

Fig. 2. Increase in the ^{13}C-content in breath-CO_2 in case of chronic calcifying pancreatitis after oral application of 300 mg benzoyl-L-tyrosyl-[1-^{13}C]alanine.

Pancreatic Function

The pancreas is an organ performing both endocrine and exocrine homeostatic function. Pancreatic insufficiency occurs in a variety of conditions, such as chronic pancreatitis, cystic fibrosis and pancreatic resection. The gold standards for estimating pancreatic exocrine function are the secretin or the secretin-pancreozymin tests. These tests can quantify the pancreatic exocrine function directly but invasively. In contrast, the ^{13}C-breath tests can estimate pancreatic exocrine function noninvasively by measuring the intraluminal activity of pancreatic enzymes. Pancreatic enzyme hydrolyses orally administered ^{13}C-labeled substrates such as triglycerides, cholesterol esters, protein and carbohydrates. The digested substrate (e.g. fatty acid) is absorbed and oxidized to $^{13}CO_2$ (fig. 1). The increase of ^{13}C-enrichment in breath after ingestion of a ^{13}C-labeled substrate reflects the intraluminal enzyme activity (fig. 2). Lipase activity has been well investigated using ^{13}C-mixed triglycerides (1,3-dist earyl-2[^{13}C]-octanoylglycerol) [13, 14], ^{13}C-triolein [15], ^{13}C-trioctanoin [2, 15] and ^{13}C-hiolein [16]. New substrates for estimating pancreatic exocrine function have been also developed in recent years.

Gastric Emptying

Gastric emptying disorder (GED) occurs in functional dyspepsia, diabetic gastroparesis, post-gastrectomy syndrome, and so on. Radioscintigraphy with test meals labeled by radioactive isotopes is the gold standard for diagnosing GED. In contrast to radioscintigraphy, ^{13}C breath tests could be exploited from the beginning of 1990s. ^{13}C-octanoic acid for solid [17] and ^{13}C-acetic acid [18] for liquid phases are employed, respectively. After passing the pylorus, the orally administered ^{13}C-labeled test meal will be absorbed and oxidized to $^{13}CO_2$ (fig. 1). In patients with diabetic gastroparesis, it is important to estimate gastric emptying because GED induces the brittleness of postprandial blood glucose levels [19]. Although the availability of ^{13}C breath tests is generally regarded for diagnosing GED, it has been pointed out that there is no standard parameter and common term in the analysis of data including the curve fitting [20].

Advantages and Disadvantages of ^{13}C

When considering the use of ^{13}C-labeled tracer, some advantages and disadvantages should be taken into account. The following eight advantages have been pointed out [21]: (1) No practical radiotracer exists for certain elements. (2) No isotope disposal costs. (3) Some studies are not practical using radiotracers. (4) Isotope effects are less than with a corresponding radiotracer. (5) Substrate content and isotope enrichment are measured simultaneously. (6) Confidence in assay specificity is very high. (7) Intramolecular location of label(s) is determined easily. (8) Simultaneous and repeated use of several tracers is possible in the same subject. In contrast, the major drawback of the ^{13}C-labeled tracer technique is the higher cost for the reagent itself and for the mass spectrometry compared with other traditional techniques. In addition, especially in breath tests, it is necessary to consider the pathway prior to exhalation of $^{13}CO_2$ including mucosal absorption, hepatic metabolism, and pulmonary function.

To the Future

The recent increase in chronic lifestyle diseases such as diabetes or obesity creates interest in the application of the ^{13}C-labeled tracer technique for the field of metabolic disease. Insulin resistance can be estimated with the ^{13}C-glucose breath test, which correlated with the gold standard diagnostic test, the hyperinsulinemic-euglycemic clamp [22, 23]. In addition, glycogen metabolism in brain [24], liver [25], and muscle [26] can be measured by using ^{13}C-NMR (nuclear magnetic resonance) in conjunction with intravenous infusions of ^{13}C-glucose. The gas chromatography-combustion-isotope ratio mass spectrometry (GC-C-IRMS) allows the elucidation of new metabolic process of lipid and fatty acid [27]. In the field of cancer research, applications of the ^{13}C-labeled tracer technique have also been reported. The high sensitivity of gas chromatography-isotope ratio mass spectrometry enabled measurement of tissue protein synthesis from small amounts of biopsy material obtained in situ [28]. ^{13}C metabolic flux analysis [29] has the potential to revolutionize cancer care to clarify metabolic fluxes in a tumor cell. New approaches with a 2-^{13}C-uracil breath test have been proposed to predict the risk of severe 5-FU toxicity in the deficiency of dihydropyrimidine dehydrogenase (DPD) [30]. Further developments in the ^{13}C-labeled tracer technique should provide a lot of additional information for clinical practice.

Conclusion

Many new research methods using ^{13}C-labeled substrates have been proposed recently. In addition to current clinical uses, such as ^{13}C-UBT, many of these new methods that have only been used in research so far could be of future use for clinical diagnosis. In order to promote these new simple techniques, we need to compare them with traditional complicated techniques and establish a standard procedure.

Acknowledgements

This work was supported in part by Grant-in-Aid for Scientific Research (C), Japan, to Yusuke Tando (KAKENHI21590869).

References

1 Lacroix M, Mosora F, Pontus M, et al: Glucose naturally labeled with carbon-13: use for metabolic studies in man. Science 1973;181:445–446.

2 Watkins JB, Schoeller DA, Klein PD, et al: ^{13}C-trioctanoin: a nonradioactive breath test to detect fat malabsorption. J Lab Clin Med 1977;90:422–430.

3 Shreeve WW, Shoop JD, Ott DG, et al.:Test for alcoholic cirrhosis by conversion of [^{14}C]- or [^{13}C]-galactose to expired CO_2. Gastroenterology 1976;71: 98–101.

4 Solomons NW, Schoeller DA, Wagonfeld JB, et al: Application of a stable isotope (^{13}C)-labeled glycocholate breath test to diagnosis of bacterial overgrowth and ileal dysfunction. J Lab Clin Med 1977; 90:431–439.

5 de Meer K, Roef MJ, Kulik W, et al: In vivo research with stable isotopes in biochemistry, nutrition and clinical medicine: an overview. Isotopes Environ Health Stud 1999;35:19–37.

6 Hepner GW: Breath analysis: gastroenterological application. Gastroenterology 1974;67:1250–1256.

7 Wolfe RR: Basic characteristics of isotopic tracers; in Wolfe RR, Chinkes DL (eds): Isotope Tracers in Metabolic Research: Principles and Practice of Kinetic Analysis, ed 2. New Jersey, Wiley, 2004, pp 1–8.

8 Furr HC, Green MH, Haskell M, et al: Stable isotope dilution techniques for assessing vitamin A status and bioefficacy of provitamin A carotenoids in humans. Public Health Nutr 2004;8:596–607.

9 Schoeller DA, Schneider JF, Solomons NW, et al: Clinical diagnosis with the stable isotope ^{13}C in CO_2 breath test; methodology and fundamental considerations. J Lab Clin Med 1977;90:412–421.

10 Peek RM Jr, Blaser MF: *Helicobacter pylori* and gastrointestinal tract adenocarcinomas. Nat Rev Cancer 2002;2:28–37.

11 Cutler AF, Havstad S, Ma CK, et al: Accuracy of invasive and noninvasive tests to diagnose *Helicobacter pylori* infection. Gastroenterology 1995;109: 136–141.

12 Chey WD, Wong BC: American college of gastroenterology guideline on the management of *Helicobacter pylori* infection. Am J Gastroenterol 2007;102: 1808–1825.

13 Vantrappen GR, Rutgeerts PJ, Ghoos YF, et al: Mixed triglyceride breath test: a noninvasive test of pancreatic lipase activity in the duodenum. Gastroenterology 1989;96:1126–1134.

14 Nakamura T, Tando Y, Kaji A, et al: $^{13}CO_2$ breath test using 13C-mixed triglycerides to evaluate pancreatic steatorrhea (in Japanese). Rinsho Shokaki Naika 2002;17:1787–1794.

15 Watkins JB, Klein PD, Schoeller DA, et al: Diagnosis and differentiation of fat malabsorption in children using ^{13}C-labeled lipids: trioctanoin, triolein, and palmitic acid breath tests. Gastroenterology 1982;82: 911–917.
16 Lembcke B, Braden B, Caspary WF: Exocrine pancreatic insufficiency: accuracy and clinical value of the uniformly labeled ^{13}C-Hiolein breath test. Gut 1996;39:668–674.
17 Ghoos YF, Maes BD, Geypens BJ, et al: Measurement of gastric emptying rate of solids by means of a carbon-labeled octanoic acid breath test. Gastroenterology 1993;104:1640–1647.
18 Braden B, Adamus S, Duan L, et al: The [^{13}C]acetate breath test accurately reflects gastric emptying of liquids in both liquid and semisolid test meals. Gastroenterology 1995;108:1048–1055.
19 Nakamura T, Takebe K, Ishii M, et al: Study of gastric emptying in patients with pancreatic diabetes (chronic pancreatitis) using acetaminophen and isotope. Acta Gastroenterol Belg 1996;59:173–177.
20 Matsubayashi T, Maruyama W: Data analysis in ^{13}C-breath test (in Japanese); in Hirano S (ed): ^{13}C-kokishiken no jissai. Tokyo, J-Way-Communications, 2001, pp 102–111.
21 Bier DM: Stable isotopes in biosciences, their measurement and models for amino acid metabolism. Eur J Pediatr 1977;156:S2–S8.
22 Lewanczuk RZ, Paty BW, Toth EL: Comparison of the [^{13}C]glucose breath test to the hyperinsulinemic-euglycemic clamp when determining insulin resistance. Diabetes Care 2004;27:441–447.
23 Mizrahi M, Lalazar G, Adar T, et al: Assessment of insulin resistance by a ^{13}C glucose breath test: a new tool for early diagnosis and follow-up of high-risk patients. Nutr J 2010;9:25.
24 Öz G, Kumar A, Rao JP, et al: Human brain glycogen metabolism during and after hypoglycemia. Diabetas 2009;58:1978–1985.
25 Roden M, Petersen KF, Shulman GI: Nuclear magnetic resonance studies of hepatic glucose metabolism in humans. Recent Prog Horm Res 2001;56: 219–237.
26 Krssak M Petersen KF, Bergeron R, et al: Intramuscular glycogen and intramyocellular lipid utilization during prolonged exercise and recovery in man: a ^{13}C and ^{1}H nuclear magnetic resonance spectroscopy study. J Clin Endocrinol Metab 2000; 85:748–754.
27 Demmelmair H, Sauerwald T, Koletzko B, et al: New insight into lipid and fatty acid metabolism via stable isotopes. Eur J Pediatr 1997;156:S70–S74.
28 Hartl WH, Demmelmair H, Jauch KW, et al: Determination of protein synthesis in human rectal cancer in situ by continuous l-^{13}C-leucine infusion. Am J Physiol 1997;272:E796–E802.
29 Metallo CM, Walther JL, Stephanopoulos G: Evaluation of ^{13}C isotopic tracers for metabolic flux analysis in mammalian cells. J Biotech 2009;144:167–174.
30 Mattison LK, Ezzeldin H, Carpenter M, et al: Rapid identification of dihydropyrimidine dehydrogenase deficiency by using a novel 2-^{13}C-uracil breath test. Clin Cancer Res 2004;10:2652–2658.

Yusuke Tando, MD, PhD
Department of Endocrinology and Metabolism
Hirosaki University School of Medicine
Honcho 53, Hirosaki, Aomori, 036 8563 (Japan)
Tel. +81 172 39 5062, Fax +81 172 39 5063, E-Mail ytando@cc.hirosaki-u.ac.jp

Yoshikawa T, Naito Y (eds): Gas Biology Research in Clinical Practice.
Basel, Karger, 2011, pp 119–124

Acetone Response during Graded and Prolonged Exercise

H. Sasaki[a] · S. Ishikawa[a] · H. Ueda[b] · Y. Kimura[c]

[a]Laboratory for Human Performance, Department of Sports Management and Science, School of Human Science, Osaka International University, [b]Laboratory Mitleben, and [c]Center for Health Science, Kansai Medical University, Osaka, Japan

Abstract
Ketone bodies consisting of acetoacetate, 3-hydroxybutirate and acetone are synthesized from fatty acids in the liver and released into the circulation. A portion of acetone from the circulation is then expired into the air. The purpose of this study was to determine the response of acetone in expired air during graded and prolonged exercises and the relationship between acetone and other respiratory parameters. Twelve female college students carried out graded and prolonged exercises during which expired gas was collected and acetone, oxygen intake and carbon dioxide excretion were analyzed. In the graded exercises, the acetone began to increase at about 35% $\dot{V}O_{2max}$, which followed the maximal fat oxidation rate at about 40% $\dot{V}O_{2max}$ and pulmonary ventilatory threshold (VT) at 46% $\dot{V}O_{2max}$. Significant correlations were found between intensity of the acetone threshold and those of the maximal fat oxidation rate ($r = 0.667$, $p < 0.05$) and the VT ($r = 0.871$, $p < 0.001$). In prolonged exercise, the acetone and the fat oxidation rate increased gradually and a logarithmically significant correlation was found ($r = 0.883$, $p < 0.001$). Therefore, the acetone in expired air increases from a relatively lower intensity during graded exercise and does so gradually during prolonged exercise. The intensity increasing the acetone correlates significantly with those that induce maximal fat oxidation rate and ventilatory threshold. The acetone level similarly correlates with the fat oxidation rate during prolonged exercise.

Lipid metabolism plays important roles for diabetes, starvation [6, 7, 9, 10] and prolonged exercise in which the role of free fatty acid (FFA) in blood is studied exclusively; however, that of ketone bodies (acetoacetate, 3-hydroxybutirate and acetone) are not so much as FFA [4, 8]. Ketone bodies are generated from fatty acids (FA) in the liver. Fatty acids partially oxidized in the liver generate 3-hydroxy-3-methylglutoryl-CoA (HMG CoA) and hydrogen ions (H^+) [6, 10]; furthermore, the former generates AcAc and the latter is released into the circulation. The AcAc is converted into 3-OBH by reduction and acetone by spontaneous decarboxylation [7, 10]. These ketone bodies,

Table 1. Physical characteristics of the subjects

	Age, years	Height, cm	Weight, kg	Body fat, %
Mean ± SD	17.3±7.3	137.6±58.6	45.2±18.3	18.4±6.8

which are strongly acidic and water-soluble, are easily released into plasma and reach tissues, while a portion of the acetone is expired into the air [5]. The acetone level in expired air correlates with plasma AcAc, 3-OHB [9] and probably FA concentrations because FA mobilized from adipose tissue increase blood ketone body levels [5]. The acetone will be an indicator for ketoacidosis [9]. Therefore, ketone bodies seem to play important roles for lipid metabolism during exercise, which may be reflected by acetone in expired air.

The purpose of this study was to investigate the response of acetone in expired air during incremental work-load testing and prolonged exercise, and to describe an examination of the relationship between the acetone and other respiratory parameters.

Materials and Methods

Subjects

Twelve female college students volunteered for this study (table 1). Each subject gave written informed consent before participating in this study, which was approved by the Ethics Advisory Committee at our university.

Procedure and Measurements

The subjects visited our laboratory twice after 12 h of fasting to avoid any effect arising from their diets. They rested in a sitting position for 5 min with electrodes mounted on their chests for ECG monitoring and recording and were wearing gas masks to collect the expired air. Thereafter, they exercised on a bicycle ergometer (Monark 818) starting from warm-up at 0 W for 3 min and continuing with the workload increasing by 17 W every minute until exhaustion for trained or near exhaustion for untrained subjects. The pedal frequency of the bicycle ergometer was stabilized at 70 revolutions per minute. ECG was continuously monitored and recorded during the final 15 s of rest, warm-up and each stage of the exercise testing. Expired air was collected into Douglas bags during each period.

About 1 week after the graded testing, each subject carried out walking or running for 2 h on a treadmill because all the subjects were not accustomed to prolonged bicycling but were to walking and running. The intensity corresponded to the heart rate inducing the maximal fat oxidation rate found during the preceding graded testing. When the heart rate attained a steady state after 5–10 min of exercise, the velocity of the treadmill was fixed and the exercise was continued. Heart rate was monitored and expired air was similarly collected into Douglas bags for 2–3 min every 15 min during the prolonged exercise.

Analysis

A portion of the expired air (250 ml) in the Douglas bags was transferred into a sampling tube in order to analyze for acetone concentration. The sampling tube was placed in an incubator kept at a constant temperature of 40°C for 10–20 min to eliminate vapor. Thereafter, a portion of expired air (2.5 ml) in the sampling tube was drawn into a syringe, which was injected into a gas chromatograph (Biogas Acetone analyzer, BAS 2000, Osaka, Japan) calibrated with a known concentration of acetone gas. The acetone concentration was automatically calculated in ppm from the area of the acetone peak indicated on the display, with a retention time was of 7.5 min. The quantity of the rest of the expired air in the Douglas bags was measured using a dry gas meter and a portion was directed into a gas analyzer (Respina IH26, SAN-EI, Tokyo, Japan) calibrated with known concentrations of O_2 and CO_2 gas to analyze for O_2 and CO_2 concentrations. Oxygen intake ($\dot{V}O_2$), carbon dioxide excretion ($\dot{V}CO_2$) and respiratory exchange ratio (R) were calculated, and, thereby, the fat oxidation rate was similarly calculated using the caloric equivalent of R and the percent kilocalories from fat with the assumption that the urinary nitrogen rate was negligible [1]. Maximal oxygen intake ($\dot{V}O_{2max}$) was determined from the regression line of heart rate and $\dot{V}O_2$. The acetone concentration, pulmonary ventilation ($\dot{V}E$) and $\dot{V}E/\dot{V}CO_2$ values of each subject were plotted against the work intensity so that those changing points were determined by the 3 students. Ventilatory threshold (VT) was determined from the initial increasing point of the $\dot{V}E$ value.

Statistics

Values obtained from the measurements and analyses were presented as means ± SD. Differences in the values were tested by analysis of variance (ANOVA). Fisher's post-hoc analysis was applied when applicable. The relationships between the acetone and other respiratory parameters were analyzed with Pearson's coefficient factor. Significance was recognized at $p < 0.05$.

Results

Predicted $\dot{V}O_{2max}$ was 44.7 ± 7.7 (33.1–59.8) ml/kg/min. The acetone concentration in expired air at rest was 0.18 ± 0.18 (0.04–0.49) ppm. During the incremental bicycle exercise testing, the acetone remained at the resting level until an intensity of 34.3 ± 5.5% $\dot{V}O_{2max}$ (0.19 ± 0.19, 0.03–0.68 ppm), but thereafter it increased in a linear fashion (fig. 1). The increase in $\dot{V}E$ value facilitated initially at an intensity of 45.8 ± 5.9% $\dot{V}O_{2max}$ (fig. 1). Intensity at which the acetone level began to increase was significantly lower than for that where the $\dot{V}E$ value was beginning to increase ($p < 0.01$) and it correlated significantly ($r = 0.871$, $p < 0.001$). The fat oxidation rate showed as a parabola against the exercise intensity and the peak value was found at an intensity of 39.6 ± 6.0% $\dot{V}O_{2max}$ which correlated significantly with the intensity of the acetone threshold ($r = 0.667$, $p < 0.05$). During prolonged exercise, acetone level and the fat

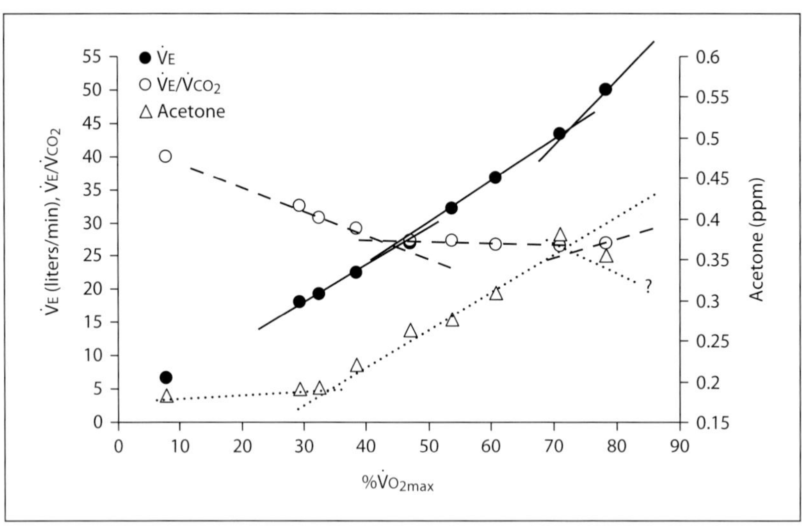

Fig. 1. Responses of acetone in expired air, pulmonary ventilation ($\dot{V}_E$) and $\dot{V}_E/\dot{V}_{CO_2}$ during graded exercise testing.

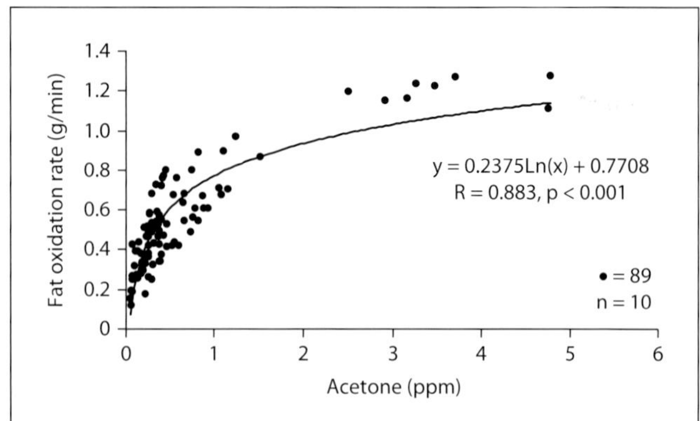

Fig. 2. Relationship between the acetone in expired air and the fat oxidation rate during prolonged exercise.

oxidation rate increased gradually and a logarithmic significant correlation was found ($r = 0.883$; fig. 2).

Discussion

This study demonstrated that acetone level in expired air increased from an intensity of 34.3% $\dot{V}O_{2max}$ during graded exercise, which followed the facilitated increase in V_E at an intensity of 45.8% $\dot{V}O_{2max}$.

Anaerobic threshold is a well-established phenomenon in which lactate from working muscles leaks into the circulation and causes metabolic acidosis, which is buffered and stimulates the respiratory center so that the parameters of pulmonary ventilation are influenced to compensate for metabolic acidosis [12]. However, previous studies [3, 11] have also pointed out that the initial changing point in the parameters preceded the increase in blood lactate from the resting level, suggesting that the other metabolic acids are buffered [3]. VT is the initial increasing point of the V_E value and regarded as the intensity compensating for metabolic acidosis. The acetone in expired air, reflecting ketone bodies in the circulation [9], increased prior to increase in the V_E value, suggesting the possibility that the ketoacidosis induced by ketone bodies contributes to respiratory compensation. Therefore, this study may indicate that ketone bodies are generated from the early stage of the exercise and strongly acidic ketone bodies and hydrogen ions (H^+) released from the liver into the circulation will be buffered and stimulate the respiratory center.

The fat oxidation rate showed a parabolic response during graded exercise and the peak value was detected at an intensity of 39.6% $\dot{V}O_{2max}$, which was partly in agreement with the previous study. Achten and Jeukendrup [1] reported that fat oxidation rate formed a parabola against the intensity, the peak value of which was detected at an intensity of 63% $\dot{V}O_{2max}$ in endurance-trained individuals.

The relative intensity inducing the maximal fat oxidation in this study was much lower than that in the previous study [1]. An increase in blood lactate inhibits fat mobilization from adipose tissue and the utilization in working muscles [2]. Furthermore, it is well known that the exercise intensity inducing the increase in blood lactate from a resting level in the trained individuals is higher than that in the untrained ones [13]. The discrepancy of the relative exercise intensity inducing maximal fat oxidation rate resulted from the training level of the subjects. Another explanation for the discrepancy may be the duration of each stage in the graded exercise. The workload was increased by 17 W for every minute in this study, while it was 35 W every 3 min in the previous study [1].

This study also demonstrated that the acetone level in expired air began to increase, which followed the maximal fat oxidation rate at the intensity of 39.6% $\dot{V}O_{2max}$. The acetone level increased gradually and correlated with the fat oxidation rate during prolonged exercise.

It is known that FFA and ketone bodies in blood increase gradually during prolonged exercise and the elevated mobilization results in increasing the utilization [4, 8]. Water-soluble ketone bodies are easily released into plasma and reach the tissues [5]. The acetone level in expired air correlates with the concentrations of plasma AcAc, 3-OHB [9] and probably FA. Ketone bodies, as described above, would be generated at an early stage of the exercise and are rapidly released. The intensity at which the acetone level in expired air began to increase correlated significantly with that which induced the maximal fat combustion rate. Therefore, this study may indicate that fat oxidation including ketone bodies is elicited during graded and prolonged exercise. However, this study cannot clarify which ketone bodies or FFA are utilized predominantly during the exercises.

Conclusions

Acetone in expired air increases from relatively lower intensity during graded exercise and does so gradually during prolonged exercise. The intensity at which the acetone level begins to increase correlates significantly with those levels that induce maximal fat oxidation rate and ventilatory threshold during graded exercise. Furthermore, the acetone level similarly correlates significantly with the fat oxidation rate during prolonged exercise.

References

1 Achten J, Jeukendrup AE: Maximal fat oxidation during exercise in men. Int J Sports Med 2003; 24:603–607.

2 Issekutz Jr B, Shaw WAS, Issekutz TB: Effect of lactate on FFA and glycerol turnover in resting and exercise dogs. J Appl Physiol 1975;39:349–353.

3 Ivy J, Costill DL, Essig D, Lower R, van Handel P: The relationship of blood lactate to anaerobic threshold and hyperventilation. Med Sci Sports 1979;11:96–97.

4 Loy SF, Conlee RK, Winder WW, Nelson AG, Arnall DA, Fisher AG: Effect of 24-h fast on cycling endurance time at two different intensities. J Appl Physiol 1986;61:654–659.

5 Newsholme EA, Satart C (eds): Regulation in Metabolism. London, Wiley, 1973.

6 Owen OE, Schramm VL: Lipid metabolism during starvation hepatic energy balance and ketogenesis. Biochem Soc Trans 1981;9:342–344.

7 Owen OE, Trapp VE, Skutches CL, Mozzoli MA, Hoeldtke RD, Borden G, Reichard GA: Acetone metabolism during ketoacidosis. Diabetes 1982;31: 242–248.

8 Ravussin E, Bogardus C, Scheidegger K, Lagrange B, Horton ED, Horton ES: Effect of elevated FFA on carbohydrate and lipid oxidation during prolonged exercise in humans. J Appl Physiol 1986;61:893–900.

9 Reichard Jr GA, Skutches CL, Hoeldtke RD, Owen OE: Acetone metabolism in humans during diabetic ketoacidosis. Diabetes 1986;35:668–674.

10 Robinsson AM, Williamd DH: Physiological roles of ketone bodies as substrates and signals mammalian tissue. Physiol Rev 1980;60:143–148.

11 Simon JS, Young JL, Gutin B, Bloom DK, Case RB: Lactate accumulation and respiratory compensation thresholds. J Appl Physiol Resp Environ Exerc Physiol 1983;54:13–17.

12 Skinner JS, McLellan TH: The transition from aerobic to anaerobic metabolism. Res Q 1980;51:234–248.

13 Williams CG, Wyndham CH, Kok R, von Rahden MJE: Effect of training on maximum oxygen intake and anaerobic metabolism in man. Int Z Angew Physiol Arbeitsphysiol 1967;24:18–23.

H. Sasaki
Laboratory for Human Performance
Department of Sports Management and Science
School of Human Science, Osaka International University
6–21–57 Tohda, Moriguchi
Osaka 570 8555 (Japan)
E-Mail hsasaki@hus.oiu.ac.jp

Yoshikawa T, Naito Y (eds): Gas Biology Research in Clinical Practice.
Basel, Karger, 2011, pp 125–132

Findings of Skin Gases and Their Possibilities in Healthcare Monitoring

Takao Tsuda[a] · Tetsuo Ohkuwa[b] · Hiroshi Itoh[b]

[a]Pico-Device. Co Ltd. and [b]Department of Material Science and Engineering, Nagoya Institute of Technology, Nagoya, Japan

Abstract

We have found several gases emanating from human skin, which we refer to as 'skin gas'. We collected skin gases from the hands, arms, fingers and other local areas on the human body, and then concentrated them in a cold trap of a 20-cm-long stainless-steel capillary, developed by us, and determined the skin gas components mostly by gas chromatography or gas chromatography/mass spectrometry with online analysis. The skin gases identified are acetone, ethanol, hydrogen, ammonia, methane, nitrogen monoxide, carbon monoxide, rose flavor after intake of essential oils, allyl methyl sulfide after intake of garlic, etc. The concentration of the components in skin gas has a good relationship with to that in blood and in some cases in breath. The variation in levels of ammonia in both skin gas and blood after the intake of protein was examined. It is clear that both the emanating rate of ammonia in the skin gas and the concentration in the blood increased after protein intake, and ammonia concentration reached its maximum level after about 2 h. The acetone concentration in skin gas correlated with the blood β-hydroxybutyrate concentration ($r = 0.669$). The skin gas acetone concentration was high (940 ppb) in a patient with diabetic ketoacidosis, and it fell to 80 ppb after insulin therapy. Skin gas ethanol was always observed even in healthy subjects without any intake of alcohol.

Noninvasive clinical monitoring is quite important in human healthcare. Breath, sweat, and saliva are used for such monitoring.

We have found several gases that emanate from human skin, which we refer to as 'skin gas'. As the chemical components in breath and sweat relate to some diseases, skin gas components also have a relationship with the state of body conditions. They can be used to monitor diseases such as diabetes. Since the components in skin gas are generally at quite low concentrations, it is necessary to use a highly sensitive analytical procedure to detect them.

We collected skin gases from the hands, arms, fingers and other local areas of the human body using a small-sized cold trap, which we developed especially for

Table 1. Relationship of skin gas and disease/condition

Skin gas components	Disease/condition	Ref.
Acetone	diabetes, ketogenetic conditions, fatness	1–3, 15
Ammonia	chronic hepatitis	7
Ethanol	alcoholic	4, 6
Hydrogen gas, methane	colonic fermentation	6, 7
Nitric oxide	state of the blood channel	9, 10
Carbon monoxide		11
Dimethyl trisulfide	fungating cancer wounds	16
nonenal	aging	17

the purpose, and determined their components mostly by gas chromatography with online analysis. We have already identified acetone [1–3], ethanol [4, 5], hydrogen [1, 6], ammonia [6, 7], methane [8], nitrogen monoxide [9, 10], carbon monoxide [11], rose flavor after intake of rose essential oils [12], allyl methyl sulfide after intake of garlic [13], nonenal as an identified chemical component aging, etc. To the best of our knowledge, we are the first to find these skin gases, particularly the light organic gases.

The method of collecting skin-permeable gases is quite noninvasive. By analyzing skin gases, we can obtain clinical information. Some of the skin gas components may be related directly to components in the blood. Thus, skin gas would be very useful for in-home healthcare and health management. As the skin gas emanates from the dermis, it is natural for the skin gas to have a relationship with the human body, namely, the state of body conditions and its diseases. A list of these relationships is shown in table 1.

Although these relationships between skin gases and diseases are almost the same as those of the breath, the analysis of skin gas is advantageous to that of the breath as follows: First, during collection of skin gas, the subject cannot control the amount of skin gas as it is under control of the autonervous system. Second, as skin gas is emanated from the skin, the gas is filtered through the dermal layer. Namely, skin gas itself has been cleaned during its emanation process. These two characteristics are important in considering the relationship between skin gas components and healthcare. The exact pathway of skin gas is unknown; it may come from sweat and from some holes on the skin surface or through the skin.

Acetone vapor has a relationship with diabetes and ketogenetic conditions. The concentration of nitric oxide gas, which is collected from the finger, wrist or hand,

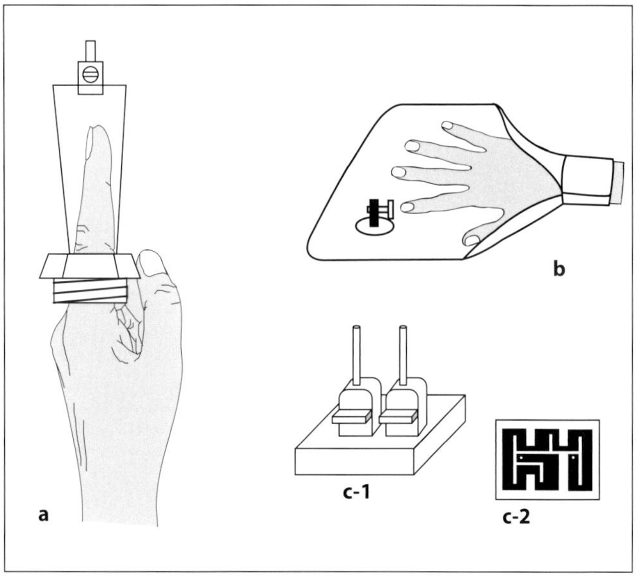

Fig. 1. Schematic design of finger (**a**) and hand (**b**) installations for skin gas collection, and a probe for the arm or flat part of the body. Bags were made with a sheet of polytetrafluoroethylene, and the gas volume was 20–100 ml for finger and 100–200 ml for hand. A probe (**c**) was carved with a whirlpool pattern on its base (c-2), and it covers 500 mm^2 of skin surface and has a volume of 0.57 ml [1, 7].

for 30 s to 3 min, is several tens of ppb, and it is possible to detect with a chemiluminescence detector. The concentration of nitric oxide would be proportional to blood pressure.

In the present paper, we describe the skin-gas-collecting devices, the concentration of each skin gas component, and its relationship with those in the breath and blood. Further, the application for diabetes and other diseases is discussed.

Analytical Procedure

Collection of Skin Gas

A typical collection device for collecting skin gas from a finger is shown in figure 1. A modified polytetrafluoroethylene bag was used for sampling from a finger. The finger was cleaned before skin gas collection. After washing under tap water for 15–60 s, the finger was wiped lightly with a paper. The cleaned finger was inserted into the modified bag, which was then closed with clips. After the gas inside the bag was sucked away lightly, 25–50 ml of clean gas was introduced and kept in. After a certain period, from 30 s to 5 min, the gas was transferred into a syringe or bag by using a homemade auto-sampler, and the gas was then analyzed [1–3, 5–14].

For collection of skin gas from a hand, arm or a local area from another part of the body surface, a modified polytetrafluoroethylene bag or sheet was used in same manner as described above. The period of skin gas collection, from 30 s up to 30 min, was selected according to the concentration of the skin gas components.

When we collect skin gas from a flat part, we can use a small probe, under which there is a groove, as shown in figure 1c. Ammonia gas from the wrist was collected during a sampling period of 3–10 min.

Cold Trapping System

The basic concept of cold trapping with pulsed direct electric current control (type NIT-P, Pico-Device Co.) is described in Kamei et al. [4]. An open stainless-steel tubing (20 cm long, inner volume of 50 to 100 μl), in which the inner wall was modified chemically, was used as a trap. The trap was kept inside a small cryostat (20 × 20 × 60 mm). It was precooled with a mixture of gas and liquid nitrogen for about 15 s, and then a sample gas was introduced for 15–60 s and then concentrated. Next, the stainless-steel tubing was heated by applying a number of pulsed direct electric currents. The desorbed sample components were led directly into the capillary column for analysis by a gas chromatograph/mass spectrometer (GC/MS). The temperature of the trap was estimated using the voltage output at both ends of the trap tube when a direct current (5 A) was applied for 6.5 ms at every 100 ms. Therefore, the present trapping system has a faster temperature response than thermo-couple devices.

Detection System

Skin gas acetone was detected with a flame ionization detector (FID) or by GC/MS with the EI mode. Skin gas ammonia was detected with a nitrogen selective detector (FTD, flame thermionic detector). Skin gas nitric oxide is possible to find directly with a chemiluminescence detector without a sample gas pre-concentration.

Skin Gases

Skin Gas Acetone

Skin gas acetone is one of the abundantly emanated gases. Skin gas acetone can be collected by any of the methods shown in figure 1. In the case of collection using method B in figure 1, 100 ml of nitrogen gas was introduced into a modified polytetrafluoroethylene bag and held closed for 5 min. The sample gas was then transferred into another bag, and 50 ml of it was analyzed. The variation of acetone concentration in both skin gas and breath was studied according to the

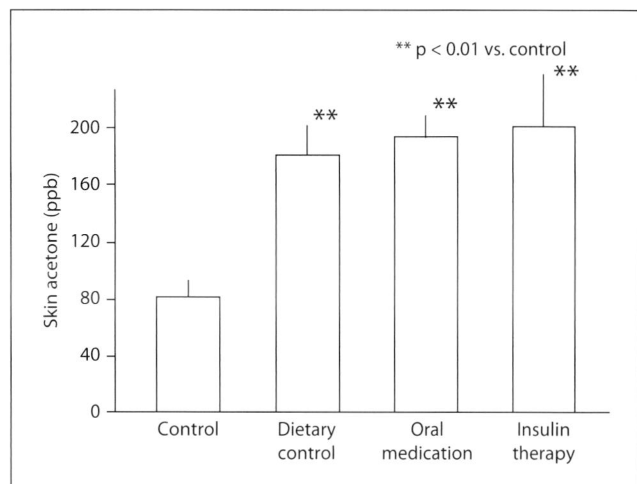

Fig. 2. Skin acetone concentration in patients with diabetes. ** $p < 0.01$ compared with control [2].

fasting period. The correlation factor between the concentrations in skin gas and in breath is 0.807 [1].

Based on the method of treatment of diabetes, subjects were classified into three groups: dietary control, treated with sulphonyurea or pioglitazone, or insulin therapy (fig. 2). Skin acetone concentrations for all the treatment groups were higher than that for the control group; however, no significant difference was observed among the treatment groups. Skin acetone concentration was correlated with blood β-hydroxybutyrate ($r = 0.669$). Skin acetone concentration was high (940 ppb) in a patient with diabetic ketoacidosis, and it fell to 80 ppb after insulin therapy [2].

Skin Gas Ammonia

The skin gas was collected from healthy subjects and subjects with hepatic diseases by using method A in figure 1. The average amounts of ammonia present in skin gas were 1.7 ± 0.4 and 2.7 ± 0.8 ng/cm^2, respectively. The results show that the ammonia was clearly present in skin gas for all subjects, and there was a significant difference between the amounts of ammonia present in healthy subjects and in those suffering from hepatic disease ($p < 0.05$) [7].

The variation in emanation rate of ammonia in skin gas collected by method C, figure 1, and the concentration of ammonia in blood after the subject's protein intake are shown in figure 3. It is clear that both the rate of ammonia in skin gas and that in the blood increased after protein intake, and reached a maximum level after about 2 h. Similar variations were recorded in the ammonia levels in skin gas and blood samples from the other 3 subjects [7].

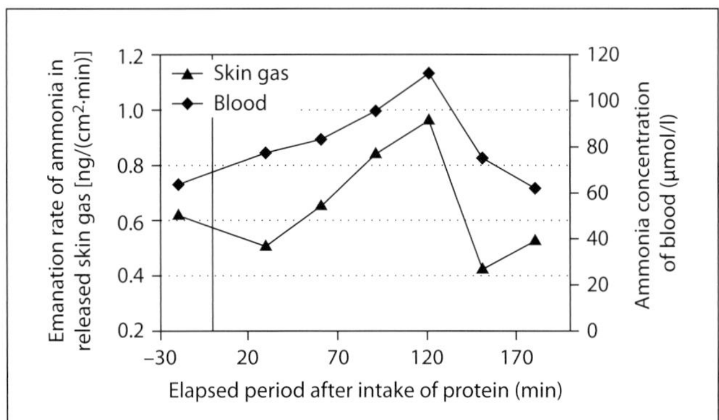

Fig. 3. Variation in amount of ammonia present in skin gases and concentration in ammonia in a blood sample from a healthy subject. Sampling period: 5 min [7].

Skin Gas Ethanol

Skin gas ethanol has a higher concentration than acetone and ammonia. Healthy people always emit alcohol from the skin even without drinking any alcohol. The average concentration of alcohol in the skin gas collected from a hand for 3 min is 0.99 ppm for 5 volunteers [5].

The variation of ethanol concentrations in skin gas and blood after the ingestion of alcohol (700 ml beer containing 5% alcohol) was examined. The values of these doses were 0.42, 0.37 and 0.51 mg/ml [4]. Ethanol concentration both in skin gas and blood is well related. Skin gas alcohol is applicable to the monitoring of alcoholism.

Skin Gas Hydrogen, Methane and Ethylene

Skin gas samples were obtained by covering a hand for 30 min with a polytetrafluoroethylene bag in which pure helium gas had been introduced. The bag, the trap system for the sample concentration, and gas chromatography were set up to avoid any contamination by air. Methane, ethylene and ethane in skin gas were successfully collected at average amounts emanated for 30 min (from 10 subjects) of 150 ± 63, 20 ± 11, and 17 ± 8 pg/cm^2 (mean ± SD), respectively [8].

Nitric Oxide and Other Gases

Nitric oxide (NO) is the most important endothelium-derived relaxing factor, which plays a pivotal role in modulating smooth muscle tone in the human conductance and resistance vessels. Ten healthy male students (21.7 ± 2.2 years; mean ± SD) volunteered as subjects. The subjects performed a repetitive wrist flexion-extension exercise at 25% maximal voluntary contraction at a pace of a time per second until voluntary exhaustion. The skin-gas samples were obtained by covering the index fingertip for 30 s with a polyfluorovinyl bag in which pure nitrogen gas was introduced, and collected in a sampling bag at rest and after exercise. The skin-gas NO concentrations significantly increased 4–4.5 min after the exercise compared to the resting values ($p < 0.05$), and the peak skin-gas NO concentrations were significantly higher than the resting values ($p < 0.01$). The blood flow levels of the annular fingertip were positively correlated with the skin gas NO concentrations of the index fingertip ($r = 0.431$, $p < 0.01$). We conclude that increased blood flow during the repetitive low-intensity exercise stimulates vascular endothelial NO production and, hence, blood flow levels correlate with skin-gas NO concentrations [9, 10].

When we separate skin gases by gas chromatography, we find a lot of chemical compounds. Zhang et al. [18] also showed that many compounds exist in skin gas, and that some of them may play an important role in detecting a disease. Further study will reveal the significance of skin gas.

Acknowledgement

This human skin gas project was supported by the Aichi Science and Technology Foundation.

References

1 Naitoh K, Tsuda T, Nose K, Kondo T, Takasu A, Hirabayashi T: New measurement of hydrogen gas and acetone vapor in gases emanating from human skin. Instrumental Sci Technol 2002;30:267–280.

2 Yamane N, Tsuda T, Nose K, Yamamoto A, Ishiguro H, Kondo T: Relationship between skin acetone and blood hydroxybutyrate concentration in diabetes. Clin Chim Acta 2006;365:325–329.

3 Mori K, Funada T, Kikuchi M, Ohkuwa T, Itoh H, Yamazaki Y, Tsuda T: Influence of dynamic handgrope exercise on acetone in gas emanating from human skin. Redox Report 2008;13:139–142.

4 Kamei T, Tsuda T, Mibu Y, Kitagawa S, Wada H, Naitoh K, Nakanishi K: Novel instrumentation for determination of ethanol concentration in human perspiration by gas chromatography and a good interrelationship in ethanol concentration in sweat and blood. Anal Chim Acta 1998;365:259–266.

5 Tsuda T: Determination of alcohol in skin gas; in Kondo T (ed): Proceedings in Biogas Analysis and Its Significance, Tokyo, 2007, p 57.

6 Kondo T, Tsuda T, Nose K, Ishiguro H, Mitsui T, Gao K, Fujiki K: Assessment of colonic fermentation by hydrogen release from human skin. Am J Gastroenterol 2002;97:1271–1272.

7 Nose K, Mizuno T, Yamane N, Kondo T, Ohtani H, Araki S, Tsuda T: Identification of ammonia in gas emanated from human skin and its correlation with that in blood, Anal Sci 2005;21:1471–1474.

8 Nose K, Nunome Y, Kondo T, Araki S, Tsuda T: Identification of gas emanated from human skin: methane, ethylene, and ethane. Anal Sci 2005;21: 625–628.

9 Ohkuwa T, Mizuno T, Kato Y, Nose K, Itoh H, Tsuda T: Effects of hypoxia on nitric oxide (NO) in skin gas and exhaled air. Int J Biomed Sci 2006;2:279–283.

10 Itoh H, Ueda A, Yoshida Y, Ohkuwa T, Yamazaki Y, Tsuda T: The relationship between blood flow and nitric oxide emanating from human skin following the wrist flexion-extension exercise. 13th Ann Congress of the European Collage of Sport Science (ECSS), Estoril, 2008.

11 Nose K, Ueda H, Ohkuwa T, Kondo T, Araki S, Ohtani H, Tsuda T: Identification and assessment of carbon monoxide in gas emanated from human skin. Chromatography 2006;27:63–65.

12 Akiyama T,Imai K, Ishida S, Itoh K, Kobayashi M, Nakamura H, Nose K, Tsuda T: Flavor components emanated from human skin after intake of essential oil. Bunseki Kagaku 2006;55:787–792.
13 Tsuda T, Nose N, Takase K, Uno A, Nakajima K: Sulfur compound emanated from human skin after intake of garlic, and suppression effect of mouse refreshing agents on it. 67th Symposium of the Japan Society of Analytical Chemistry, Akita, 2006.
14 Naitoh K, Inai Y, Hirabayashi T, Tsuda T: Direct temperature-controlled trapping system and its use for the gas chromatographic determination of organic vapor released from human skin. Anal Chem 2000;72:2797–2801.
15 Turner C, Parekh B, Walton C, Spanel P, Smith D, Evans M: An exploratory comparative study of volatile compounds in exhaled breath and emitted by skin using selected ion flow tube mass spectrometry. Rapid Commun Mass Spectrom 2008;22:526–532.
16 Shirasu M, Nagai S, Hayashi R, Ochiai A, Touhara K: Dimethyl trisulfide as a characteristic order associated with fungating cancer wounds. Biosci, Biotechnol Biochem 2009;73:2117–2120.
17 Haze S, Gozu Y, Nakamura S, Kohno Y, Sawano K, Ohta H, Yamazaki K: 2-Nonenal newly found in human body order tends to increase with aging. J Invest Dermatol 2002;116:520–524.
18 Zhang ZM, Cai JJ, Raum GH, Li GK: The study of fingerprint characteristics of the emanations from human arm skin using the original sampling system by SPME-GC/MS. J Chromatogr 2005;822:244–252.

Takao Tsuda
Pico-Device Co., Nagoya Ikourennkei Incubator
Chikusa 2–22–8
Chikusa-ku, Nagoya 464-0858 (Japan)
Tel. +81 52 735 7327, Fax +81 52 526 8616, E-Mail tsuda@pico-device.co.jp

Yoshikawa T, Naito Y (eds): Gas Biology Research in Clinical Practice.
Basel, Karger, 2011, pp 133–143

Phytoncide – Its Properties and Applications in Practical Use

Masato Nomura

Department of Biotechnology and Chemistry, Faculty of Engineering, Kinki University, Hiroshima, Japan

Abstract

Lately, terms such as 'free radical' and 'active enzyme' have been used frequently in topics related to health and illness. Under normal conditions, free radicals and active enzymes that are produced within an organism are erased by the antioxidant enzyme SOD (superoxide dismutase). However, under excessive oxidant stress, the biological defense mechanisms become overwhelmed and the role of antioxidants becomes important. This research focuses on the relationship between forest therapy and phytoncide solution (4 varieties), a concentrated volatile component released from trees and plants. The antioxidant potential (via a DPPH radical scavenging effect experiment and active oxygen inhibition experiment), bacterial eradication and deodorization effects of phytoncide and its ability to relieve and reduce restraint stress will be examined. In the DPPH radical scavenging experiment, all four types of phytoncide solution (concentrate) demonstrated a radical scavenging rate that was over 97% and higher than that of the reference material, α-tocopherol (95.0%). Furthermore, the D-type liquid had a 50% radical scavenging concentration (SC_{50}) equal to that of α-tocopherol and proved to possess the highest radical scavenging effect observed. In the active oxygen scavenging experiment, although α-tocopherol, with its ability to inhibit oxidation of unsaturated fatty acid in phosphatide forming the plasma membrane, maintain biomembranes and scavenge free radicals, is a most effective antioxidant agent, an active oxygen scavenging effect for α-tocopherol was not observed. Therefore, an active oxygen scavenging experiment was done on the four types of phytoncide solution, and the AB-type liquid with high bactericidal capacity as the main component was found to show a good active oxygen scavenging rate (60.7%). Because traditional and natural images project a positive image to consumers in the cosmetic market, plant-derived compounds are actively used. Ordinary hand-washing eradicates transient flora and hygienic hand-washing eradicates resident flora. In a test of bacterial eradication during daily activities from the morning commute until lunch time, hand-washing using liquid soaps with phytoncide (AB-type liquid) demonstrated higher eradication effects toward resident microbiota, i.e. the long bacteria that form velutinous colonies, than ordinary liquid soaps. In addition, hand-washing with phytoncide resulted in a remarkable decrease in the number of short bacteria that remained after washing. The four representative elements of bad odor are ammonia, hydrogen sulphide, trimethylamine and methyl mercaptan, and the majority of a bad odor results from the foul odor caused by oxidation. The deodorization ability of phytoncide derives from chemical neutralization, and degradation and detoxification of malodorous components, and it brings a refreshing effect by possessing an odor similar to air in the forest. Therefore, the use of phytoncide will be beneficial in maintaining and controlling a pleasant environment in institutions. As phytoncide is used in aroma therapy, the components of phytoncide are considered to approximate to essential oils. Phytoncide mist does not cause a negative reaction in the circulatory dynamics of normally bred, spontaneously hypertensive rats, and the stress-reducing effect of phytoncide mist toward the highly stress-sensitive disease model animal, the stroke-prone spontaneously hypertensive rat, was examined by monitoring heart rate, systolic blood pressure and diastolic blood pressure. These experiments indicate that under a stressful environment, inhalation of phytoncide has the function of restraining activation of the circular and sympathetic nervous system, thereby reducing stress and reforming mental function. In this research, phytoncide solution derived from a combination of 118 kinds of plants was assessed in terms of its antioxidant potential, sterilization, odor-eliminating and stress-reducing properties. Clearly, the bioactivity of phytoncide solution, including its terpenoid components, is related to these effects.

In highly information-oriented modern urban societies, people live under severe stress, and health disorders caused by stress have become a serious problem [1, 2]. High pressure on the body due to stress causes disequilibrium in the body and results in various problems both physically and mentally. One of the gravest problems is that 80% of lifestyle-related diseases such as hypertension, cardiac infarction, diabetes and stomach ulcer are caused by stress [3, 4]. Under such conditions, people seek spiritual comfort, and awareness of the ability of natural odors to relieve stress (aromatherapy) is rising year by year.

'Forest therapy' is a term created by the forestry agency 25 years ago and has been under the spotlight as a potential recreation and health management method. The 'forest therapy' effect derives from a volatile component, phytoncide, that is secreted by and emanates from the forest; phytoncide activates the body and results in beneficial health effects [5, 6]. The main volatile content of phytoncide are terpenoids, which are known to possess bactericidal, antiviral, disinfection and antibacterial properties [7–10]. In ancient Greece and Rome, plant-derived essences were used in medical treatment and especially as antibacterial substances [11, 12].

Lately, terms such as 'free radical' and 'active enzyme' have been used frequently in topics related to health and illness. An active enzyme is an enzyme that is taken in and causes alterations to substances in the body, thereby producing superoxide, hydroxyl radical, hydrogen peroxide and singlet oxygen species. Free radicals and active enzymes cause physical damage, including denaturation of protein, lipid oxidation deactivation of enzymes and DNA incision that can harm biomembranes and genes, and cause disorders such as aging, arteriosclerosis, diabetes and cancer [13].

Under normal conditions, free radicals and active enzymes that are produced within an organism are erased by the antioxidant enzyme SOD (superoxide dismutase). However, under excessive oxidant stress, the biological defense mechanisms become overwhelmed and the role of antioxidants becomes important [14, 15].

This research focuses on the relationship between forest therapy and phytoncide solution (4 varieties), a concentrated volatile component released from trees and plants. The antioxidant potential (via a DPPH radical scavenging effect experiment and active oxygen inhibition experiment) [16], bacterial eradication and deodorization [17–19] effects of phytoncide, and its ability to relieve and reduce restraint stress [20, 21], will be examined.

Preparation of Phytoncide Solution

Four varieties of phytoncides solution were prepared as follows: a type liquid, comprising plant extract of trees as the main constituent; AB-type liquid, comprising plant extract with high bactericidal activity as the main constituent; CY-type liquid, comprising plant extract without allergen reaction, as the main constituent and D-type liquid, comprising plant extract of plants, as the main constituent. All of the phytoncide solutions were blends of essences combined from 118 kinds of plants, as indicated [16].

Antioxidant Activity

The volatile compounds of phytoncide solution are low molecular weight organic compounds called terpenoids, and mainly comprise *d*-limonene, (+)-fenchone, (+)-camphor, (+)-linalool, (+)-fenchol, β-cedrene, α-cedrene, (+)-α-terpineol, (+)-α-pinene, (+)-β-pinene, (+)-β-citoronellol, and anethole [16, 19].

In the DPPH radical scavenging experiment, all four types of phytoncide solution (concentrate) demonstrated a radical scavenging rate that was over 97% and higher than that of the reference material, α-tocopherol (95%). Furthermore, the D-type liquid had a 50% radical scavenging concentration (SC_{50}) equal to that of α-tocopherol and proved to possess the highest radical scavenging effect observed. From a component search (ether extract) of the D-type liquid, 27 chemical structures (including the main terpenoids, linalool, thujopsene and α-terpinol) were identified. In addition, the oil content was fractionated into a hexane extraction layer and a water layer, and each was tested for its DPPH radical scavenging ability. As a result, high radical scavenging ability was observed in the water layer, and 15 phenylpropanoids, as indicated in figure 1, were identified from a component search. Among the 15 phenylpropanoids, the DPPH radical scavenging ability of 8 compounds was examined, and 2-methoxyphenol, 2-methoxy-5-methylphenol, 2-methoxy-4-methylphenol, eugenol and 4-allyl-2,6-dimethoxyphenol demonstrated a radical scavenging rate (over 90%) higher than that of α-tocopherol. At the same time, all of these compounds were found to have an SC_{50} value of 1–4 and proved to have the highest radical scavenging ability. The relationship between these compounds and their expression of activity seems to be the o-composition of hydroxyl and methoxy in the benzene ring. In addition, a high radical scavenging rate was observed in compounds with methoxy or allyl

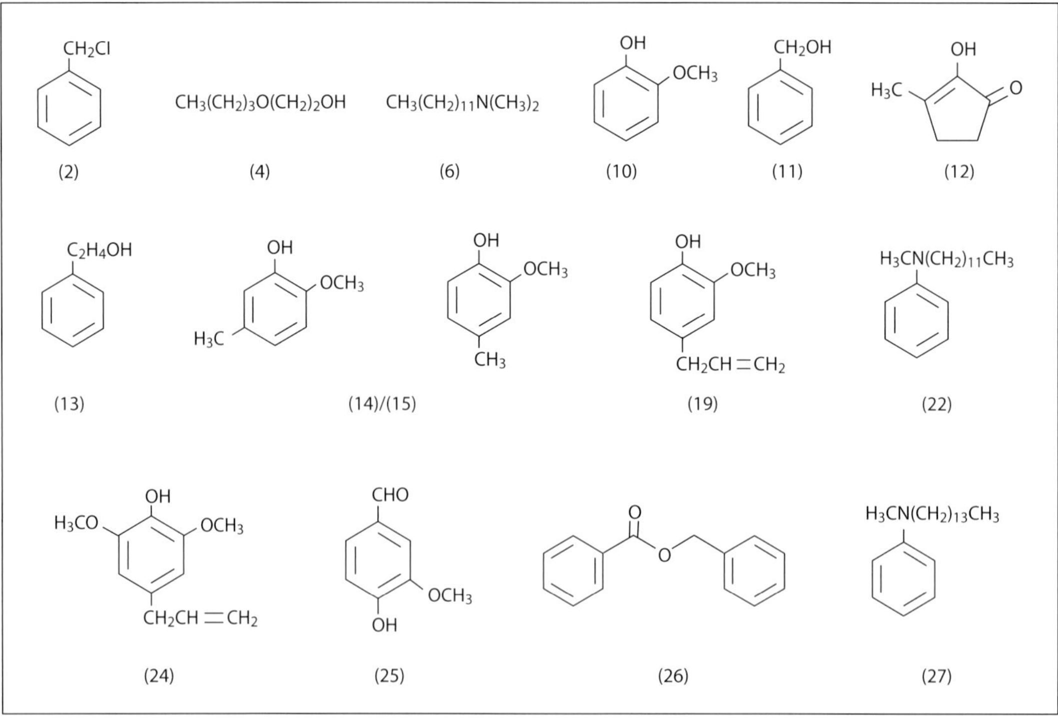

Fig. 1. Chemical structure of principal compounds in the water layer (D-type phytoncide solution).

as the third substituent. On the other hand, a combination of hydroxyl and methoxy in an m-composition compound also resulted in a high radical scavenging rate.

In the active oxygen scavenging experiment [23], although α-tocopherol, with its ability to inhibit oxidation of unsaturated fatty acid in phosphatide forming the plasma membrane, maintain biomembranes and scavenge free radicals, is a most effective antioxidant agent, an active oxygen scavenging effect for α-tocopherol was not observed. Therefore, an active oxygen scavenging experiment was done on the four types of phytoncide solution, and the AB-type liquid with high bactericidal capacity as the main component was found to show a good active oxygen scavenging rate (60.7%). In addition, the A-type and CY-type phytoncide solutions demonstrated an active oxygen scavenging rate of around 50%.

To confirm the active oxygen scavenging ability of phytoncide solution, we used a nitric oxide inhibitory experiment [24], based on the oxidation of nitrogen dioxide to nitrogen monoxide. In this inhibitory experiment, the amount of nitrogen monoxide produced from different concentrations (1.0–0.25%) of phytoncide solution were measured. As a result, as table 1 indicates, differences in the concentration of the A type and AB type phytoncide solutions derived distinguishable differences in inhibitory effect.

Table 1. Nitrogen monoxide inhibition and lipid peroxide inhibition by phytoncide solution

Phytoncide	Nitric oxide inhibitive rate, %[1]			Lipid peroxide inhibitive rate, %[2]		
	1.0[c)]	0.5	0.25	1.0[3]	0.5	0.25
A-type	70.3	45.1	31.9	57.8	33.9	17.5
AB-type	72.1	55.8	39.6	52.4	27.3	11.5
CY-type	57.2	44.1	35.4	43.8	26.8	15.5
D-type	47.4	34.5	27.4	61.3	35.8	21.8

[1] $NaNO_2$ solution 50 μM.
[2] MDA standard 2.0 μM.
[3] Concentration (mg/ml).

In general, lipid peroxide is formed by reacting active oxygen and lipid; therefore, under the assumption that one of the functions of phytoncide solution, namely its active oxygen effect, engages in these reactions and inhibits lipid peroxide formation, an oxidized lipid inhibitory experiment [25] was conducted. In this experiment, dependence on the concentration of phytoncide solution (1.0–0.125%) was high. For example, a 1.0% concentration of A-type and D-type liquid resulted in an inhibition of 58–61% of oxidized lipid, whereas at low concentration (0.125%) this rate decreased to 11–12%. In summary, these inhibitory experiments show that phytoncide solution, which demonstrates the forest therapy effect, prevents a rise in active oxygen at minimum and also eliminates active oxygen; these functions lead to the prevention of oxidation in the body.

Bacterial Eradication and Deodorization

Effectiveness as a Liquid Soap

Because traditional and natural images project a positive image to consumers in the cosmetic market, plant-derived compounds are actively used. Incorporating essences extracted from familiar plants into cosmetics has the advantage of creating people-friendly products. Among the phytoncide solutions containing various odor components, the AB-type liquid formed by plant essences with high bactericidal properties has a content of around 50% terpenoids such as guaiacol, α-n-methyl ionone and cedrol, in addition to several kinds of phenylpropanoid (benzyl alcohol, β-phenethyl alcohol, o-cresol and 4-ethyl phenol) and fatty acid (octanoic acid, decanoic acid, dodecanoic acid and hexadecanoic acid). Therefore, the bacterial eradication (table 2) and

Table 2. Antimicrobial and antifungal properties of the AB- and D-type phytoncide solutions

Phytoncide	Concentration, %	Bacteria		Fungi		Yeast	
		3 days	7 days	1 week	2 weeks	1 week	2 weeks
AB-type	0.1	0[1]	0	1.0×10^3	1.5×10^3	2.5×10^1	0
	0.5	0	0	1.0×10^1	2.0×10^1	0	0
	1.0	0	0	3.0×10^1	0	0	0
D-type	0.1	1.5×10^1	1.5×10^3	6.0×10^3	4.0×10^3	3.0×10^1	0
	0.5	0	0	1.6×10^2	1.3×10^2	0	0
	1.0	0	0	1.2×10^2	4.5×10^1	0	0

[1] cfu/g.

deodorization properties of liquid soaps (lauric acid, myristic acid and palmitic acid) compounded with 1.0% of the AB-type phytoncide solution were examined [17].

Firstly, the compounded higher fatty acids were found to possess high water solubility and to excel in foaming properties and detergency. Bacterial flora are present on hand skin. Ordinary hand-washing eradicates transient flora and hygienic hand-washing eradicates resident flora. In a test of bacterial eradication during daily activities from the morning commute until lunch time, hand-washing using liquid soaps with phytoncide (AB-type liquid) demonstrated higher eradication effects toward resident microbiota, i.e. the long bacteria that form velutinous colonies, than ordinary liquid soaps. In addition, hand-washing with phytoncide resulted in a remarkable decrease in the number of short bacteria that remained after washing.

Secondly, the deodorization ability of this liquid soap, which showed notable bacterial eradication effects, was examined. The deodorization ability of the liquid soap with phytoncide was evaluated by 10 volunteers using a 5 points per person system and hand-washing (20–60 s) during daily activities, and compared to ordinary liquid soaps. As a result, the volunteers evaluated that the liquid soap with phytoncide required a shorter hand-washing time and showed good deodorization ability. In addition, for thorough hand-washing with running water for over 40 s, the evaluation standard was judged to be just under 1, and the onset of good deodorization effects was confirmed.

Deodorization Ability in the Indoor Environment

The four representative elements of bad odor are ammonia, hydrogen sulphide, trimethylamine and methyl mercaptan, and the majority of a bad odor results from the foul odor caused by oxidation. Therefore, we assessed the deodorization and refreshment ability of phytoncide (PT150 {type A: type D = 1:1}) on commonly found indoor odors, inclusive of ammonia, domestication products and other chemicals, an on the odor of animals, with respect to the environmental improvement of public institutions such as universities. An air-conditioning system was placed and functioning in each of the rooms according to the guideline standards. However, the effluvium of mixed odors in the rooms was noted by some people as unpleasant. In addition, a high concentration of formic aldehyde (sick building syndrome) in medical institutions is considered to be related to the pathogenesis of chemical sensitivity and is treated as a serious health problem [18].

The odor environment of the participating institution was evaluated as one of four levels, 'strong odor', 'some odor', 'little odor' and 'no odor', and the phytoncide deodorization effect as one of four levels, 'very effective', 'effective', 'less effective' and 'not effective'. As a result, around 80% of the bad odor environments were improved. For the deodorization of formic aldehyde, whereas the concentration of formic aldehyde with vapor spray was steady at 0.86 ± 0.03 ppm/h, the concentration with phytoncide mist decreased over time (second day, 0.61 ± 0.02 ppm/h; third day, 0.52 ± 0.03 ppm/h), and a significant difference in odor elimination was observed. Thus, spraying phytoncide mist constantly can be expected to produce higher deodorization effects. Moreover, because the conventional ways of deodorization, including physical absorption by porous, activated carbon or zeolite, and masking by other odors or odorous substance resolution (ozone) and photocatalysis (titanium oxide), contain both positive and negative points, combining phytoncide and these methods will be beneficial in bringing about new environmental improvements. The deodorization ability of phytoncide derives from chemical neutralization, and degradation and detoxification of malodorous components, and it brings a refreshing effect by possessing an odor similar to air in forest. Therefore, the use of phytoncide will be beneficial in maintaining and controlling a pleasant environment in institutions.

Effect of Phytoncide on the Body

As phytoncide is used in aroma therapy, the components of phytoncide are considered to approximate to essential oils. In animal experiments on rats and mice, a sedating effect has been reported as a result of the inhalation of essential oils including lavender and citrus [26–29]. These essential oils pass through the GABA (gamma-aminobutyric acid) system in the brain and central nerve system to produce sedating effects. A stress reaction is considered to be a reaction of the endocrine, vascular, nervous, muscular and immune systems of the body that maintains homeostasis during a disorder caused by certain stimuli.

Against an environmental stress, the nervous system autonomously controls cyclic neurogenic adjustment to maintain homeostasis and plays an important role in a cyclic reflex response, which is adjusted by a balance of the sympathetic and parasympathetic

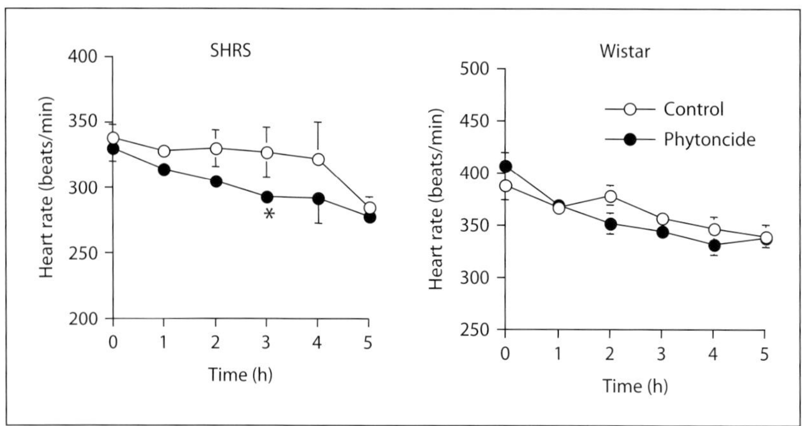

Fig. 2. Change in the heartbeat count of rats according to restriction stress load state. Data are mean ± SEM. * $p < 0.05$.

nervous systems. Because phytoncide (PT-150) mist does not cause a negative reaction in the circulatory dynamics of normally bred, spontaneously hypertensive rats, the stress-reducing effect of phytoncide (PT-150) mist toward the highly stress-sensitive disease model animal, the stroke-prone spontaneously hypertensive rat, was examined by monitoring heart rate, systolic blood pressure and diastolic blood pressure [20, 21].

Effect on Restriction Stress

SHRSP/Izm (n = 15) and Wistar (n = 12) rats were tethered by a cloth holder for the sphygmomanometry. As a result, in terms of heart rate (fig. 2), although Wistar rats did not show a significant difference, SHRSP/Izm rats showed a significant difference after three hours and a tendency for a temporal decrease as compared to the beginning of the holding. With regard to systolic blood pressure (fig. 3), SHRSP/Izm rats showed a decrease over time (1→5 h) and a significant difference was observed. Wistar rats also showed a significant difference over time (3→4 h). With regard to diastolic blood pressure (fig. 4), a significant decrease was observed in SHRSP/Izm rats after 1–3 h and in Wistar rats after 3–4 h. Furthermore, in examination of the plasma catecholamine density, a decrease of about 20% in both adrenaline and noradrenaline levels was observed in SHRSP/Izm rats.

Influence on Water Immersion Restriction Stress

After food deprivation for 24 h, the experimental SHRSP/Izm rats were placed in a constant-temperature bath (40°C) for 4 h as a test of water immersion restraint stress.

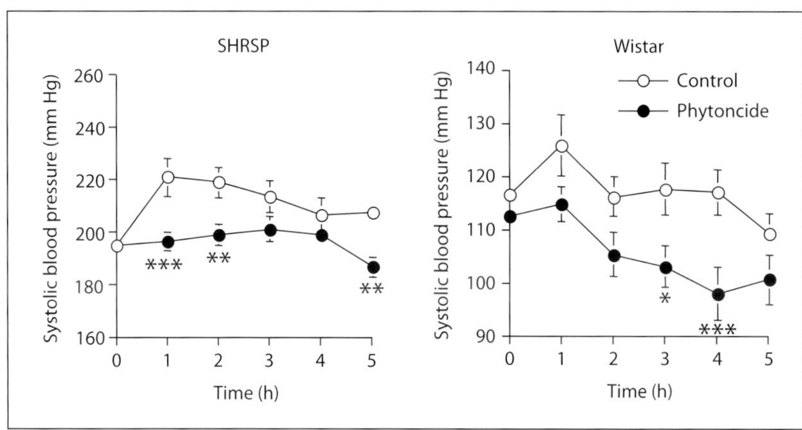

Fig. 3. Change in the systolic blood pressure of rats according to restriction stress load state. Data are mean ± SEM. * $p < 0.05$, ** $p < 0.01$, *** $p < 0.001$.

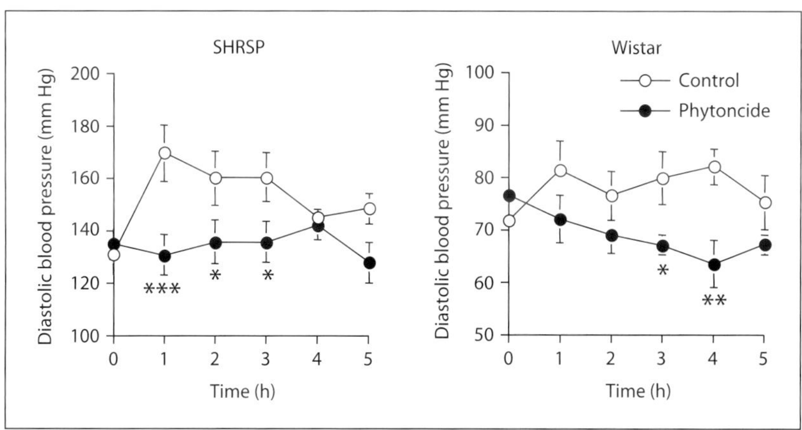

Fig. 4. Change in the diastole blood pressure of rats according to restriction stress load state. Data are mean ± SEM. * $p < 0.05$, ** $p < 0.01$, *** $p < 0.001$.

Afterwards, phytoncide was sprayed for a fixed time. Each 1 mm^2 of blood spots on the glandular stomach was counted as one and digitalized, and the dimensions of the bleed insult were measured digitally. Summation of the damage was compared as a damage coefficient. As a result, as figure 5 shows, with regard to gastric ulcers and hemorrhage, although much damage was observed in Wistar rats, this damage was significantly less in SHRSP/Izm rats. Furthermore, the number of blood spots was significantly smaller in SHRSP/Izm than in Wistar rats. Thus, these experiments indicate that under a stressful environment, inhalation of phytoncide has the function of restraining activation of the circular and sympathetic nervous system, thereby reducing stress and reforming mental function.

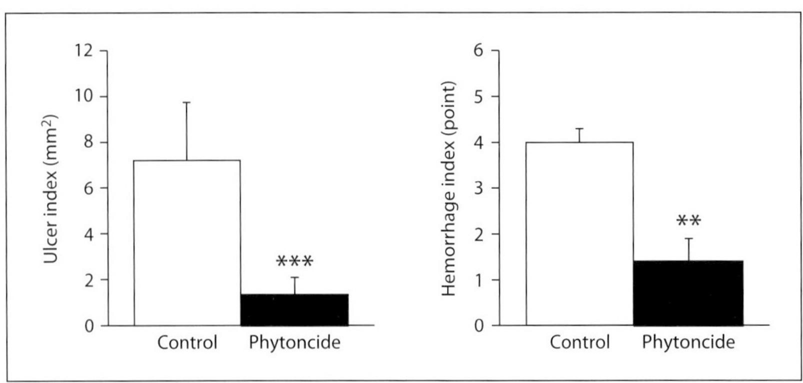

Fig. 5. Gastric ulcer and hemorrhage in SHRSP after the water immersion stress test. Data are mean ± SEM. ** $p < 0.01$, *** $p < 0.001$.

Conclusion

In this research, phytoncide solution derived from a combination of 118 kinds of plants was assessed in terms of its antioxidant potential, sterilization, odor-eliminating and stress-reducing properties. Clearly, the bioactivity of phytoncide solution, including its terpenoid components, is related to these effects. By revealing the body's systems that are used to metabolize phytoncide solution, including its terpenoid components, and their relation to antioxidation of body and reducing stress, further developments in this discipline are expected in the future. On the other hand, many fundamental investigations on the odor-eliminating effects of phytoncide solution have already been carried out. Revealing the mechanism by which phytoncide produces its odor-eliminating effects is expected to enable the creation of eco-friendly deodorants.

References

1 Tadaki Y, Kira T: Homo sapiens and Forest. Tokyo, Kyoritsu Publishers, 2002, pp 229–271.
2 Tsuchiya T, Hosoi J: Psychoneuroendocrinological and psychoneuroimmunological effects of stress on skin functions. Fragrance J 1996;24:26–34.
3 Nomura S, Naitoh H: Overcrowding stress as an animal model of depression. Jpn J Clin Psychiatry 1991;20:429–436.
4 Nishikaze O, Furuya E: Stress and anticortisols – 17-ketosteroid sulfate conjugate as a biomarker in tissue repair and recovery. J UOEH 1998;20:273–295.
5 Strobel GA: Phytotoxins produced by plant parasites. Annu Rev Plant Physiol 1974;25:541–566.
6 Nakagawa M, Nagai H: Evaluation for relieving effects to stress with odoriferous compounds. Fragrance J 1991;19:44–49.
7 Isacoff H: Aromatics as bactericides. Cosmet Toiletries 1981;96:69–76.
8 Yatagai M: Recent research and development in phytoncides. Fragrance J 1986;6:247–252.
9 Saito T: Effects of fragrances on stress alleviation and sleep. Kohshokai 1999;23:87–91.
10 Aoshima H, Hamamoto K: Potentiation of GABAA receptors expressed in *Xenopus* oocytes by perfume and phytoncide. Biosci Biotechnol Biochem 1999;63:743–748.
11 Kamiyama K: Ecology of forest and exhalating materials effect. Fragrance J 1984;65:7–11.

12 Ohtsuka Y, Yabunaka N, Takayama S: Shinrinyoko (forest-air bathing and walking) effectively decreases blood glucose levels in diabetic oatients. Int J Biometeorol 1998;41:125–127.
13 Yoshikawa T: All of antioxidation materials. Sentanigakusha 1998; pp 16–24.
14 Osawa T: New functions of foods, and analytical methods for their evaluation. Evaluation and analysis of natural antioxidants. Tech J Food Chem Chem 1988;4:36–42.
15 Osawa T: Free radicals-basis and clinics: aging study using antioxidants. New Horizon Med 1993;25:3327–3332.
16 Abe T, Nomura M: Chemical constituents of phytoncides and their antiovidative activities. Aroma Res 2006;7:56–62.
17 Abe T, Tanimoto S, Hisama M, Mihara Y, Nomura M: Antimicrobial and Deodorization Activities of phytoncide solution. Bokin Bobai 2007;35:489–495.
18 Kawakami K, Kawamoto M, Horie T, Mihara Y, Nomura M, Yamada T, Kobayashi U, Otani H: Examination of deodorant effect of phytoncide in the indoor environment. Res Soc Environ Control Tech 2006;24:17–22.
19 Abe T, Hisama M, Tanimoto S, Shibayama H, Mihara Y, Nomura M: Antioxidation effects and antimicrobial activities of phytoncide. Biocontrol Science 2008;13:23–27.
20 Kawakami K, Shimosaki S, Ishida I, Gonbi N, Nomura M, Nabika T: Effects of phytoncides on cardiovascular parameters of SHR. Kyushu Lab Anim Magazine 2006;22:23–27.
21 Kawakami K, Shimosaki S, Tongu M, Nomura M, Kobayashi Y, Nabika T, Yamada T: Influence of phytoncides on restraint stress response in the stroke-prone spontaneously hypertensive rat (SHRSP). Kyushu Lab Anim Magazine 2007;23:57–62.
22 Tominaga H, Kobayashi M, Goto T, Kasemura K, Nomura M: DPPH radical-scavenging effect of several phenylpropanoid compounds and their glycoside derivatives. YAKUGAKU ZASSHI 2005;125: 371–375.
23 Tada T, Nomura M, Shimomura K, Fijihara Y: Synthesis of karahanaenone derivatives and their inhibition properties toward tyrosinase and superoxide scavenging activity. Biosci Biotech Biochem 1996;60:1421–1424.
24 Yamamoto H: In vivo and in vitro effects of melatonin or ganglioside GT1B on L-cysteine-induced brain mitochondrial DNA damage in mice. Toxicol Sci 2003;73:416–422.
25 Delphine L: Altered vascular function in fetal programming of hypertension. Stroke 2002: pp 2992–2998.
26 Saeki Y: The effect of foot-bath with or without the essential oil of lavender on the autonomic nervous system: a randomized trial. Complement Ther Med 2000;8:2–7.
27 Edge J: A pilot study addressing the of aromatherapy massage on mood, anxiety and relaxation in adult mental health. Complementary Ther Nurs Midwifery 2003;9:90–97.
28 Kim JT, Wajda M, Cuff G, Serota D, Schlame M, Axelrod DM, Guth AA, Bekker AY: Evaluation of aromatherapy in treating postoperative pain: pilot study. Pain Practice 2006;6:273–277.
29 Suzuki A, Okubo N: Aromatherapy research in nursing and its present state. Res Rep St. Luke's Coll Nurs 2009;35:17–27.

Dr. M. Nomura
Department of Biotechnology and Chemistry
Faculty of Engineering, Kinki University
Hiroshima 739-2116 (Japan)
Tel. +81 82 434 7000, Fax 82 434 7011, E-Mail nomura@hiro.kindai.ac.jp

Author Index

Choi, K.-S. 65

Dogishi, K. 81

Hahm, K.-B. 65
Han, S.J. 65
Hashimoto, S. 56
Hayashi, S. 81
Hirata, I. 73
Honda, Y. 6

Inoue, M. 56
Ise, F. 81
Ishikawa, S. 119
Itoh, H. 125

Kabe, Y. 1
Kajimura, M. 1
Kim, E.-H. 65
Kim, J.H. 65
Kimura, Y. 119
Koyama, M. 81

Maeda, T. 6
Maruyama, J. 43
Maruyama, K. 43
Matsuhashi, Y. 112
Matsumoto, A. 112
Matsumoto, M. 6

Naito, Y. VII, 35, 73
Nakajima, H. 6
Nakamura, T. 112
Nakao, A. 15, 91
Nomura, M. 133

Ock, C.Y. 65
Ohkuwa, T. 125

Sasaki, H. 119
Sasaki, Y. 6
Sawano, M. 24
Shimada, N. 6
Shime, N. 56
Suematsu, M. 1
Sugimoto, M. 6

Takagi, T. 35, 73
Takemoto, I. 6
Takeuchi, K. 81
Tanaka, H. 112
Tando, Y. 112
Tsuda, T. 125

Uchida, M. 100
Uchiyama, K. 35
Ueda, H. 119
Urita, Y. 6

Watanabe, T. 6

Yanagimachi, M. 112
Yasuda, M. 81
Yoshikawa, T. VII, 35, 73

Zhang, E. 43

Subject Index

Acarbose, hydrogen response 97
Acetone, *see also* Ketone bodies
expiration correlation with ventilation rate and fat oxidation rate of exercise 120–124
skin gas studies 128, 129
Acute lung injury, nitric oxide therapy 49, 62
Acute respiratory distress syndrome, nitric oxide therapy 49, 62
Alzheimer's disease, hydrogen therapy 95
Ammonia, skin gas studies 129
Apoptosis, carbon monoxide inhibition 37

Bicarbonate, stimulation in duodenum by hydrogen sulfide 81–89
Bronchopulmonary dysplasia, nitric oxide for prevention in preterm infants with respiratory distress syndrome 60, 61

Cachexia, carbon-13 breath test for evaluation 106
Carbon-13 breath test
advantages and limitations 116
applications in experimental animals 100–103, 106–109
cachexia evaluation 106
clinical use prospects 116, 117
collection from animals 103, 104
gas chromatography/mass spectrometry analysis 104
gastric emptying evaluation 105, 115, 116
gastrocecal transit time evaluation 106
Helicobacter pylori infection evaluation 104, 105, 114
liver function evaluation 105
pancreatic function evaluation 105, 106, 115
pharmacokinetic data analysis 104
principles 113, 114
substrates 113
sugar and lipid metabolism evaluation 106
test meals for animal testing 103
Carbon monoxide
anti-inflammatory effect 38, 39
apoptosis inhibition 37
cystathionine β-synthase as sensor 3, 4
functional overview 35, 36
guanylate cyclase binding 3, 36
hemodynamic monitoring with expired gas analysis
advantages 31–33
carbon monoxide-saturated autologous blood preparation 25, 26
carboxyhemoglobin fraction estimation 27
cardiac output estimation 28, 29
circulating blood volume estimation 28–30
constants 27, 28
continues determinations of CO and CO_2 26
disadvantages 3
end-tidal carbon monoxide estimation 26, 27
overview 24, 25
validation 28–31
hepatic stellate cell effects 2, 3
inflammatory bowel disease treatment 38–40
potassium channel activation 36, 37
properties and therapeutic application 18, 19
releasing molecules 17

sinusoid relaxation mechanism 3
synthesis, *see* Heme oxygenase
Toll-like receptor inhibition 19
Cardiac output, estimation from carbon monoxide expired gas analysis 28, 29
Chronic obstructive pulmonary disease, nitric oxide therapy 49, 50
Circulating blood volume, estimation from carbon monoxide expired gas analysis 28–30
Cisplatin, hydrogen prevention of nephrotoxicity 95
Colitis, *see also* Inflammatory bowel disease
hydrogen prevention of dextran sodium sulfate-induced colitis 95, 96
hydrogen sulfide
inflammatory response 76–78
prospects for study 78, 79
synthesis inhibitor studies 75, 76
Congestive heart failure, nitric oxide therapy 50
Crohn's disease, *see* Inflammatory bowel disease
Cystathionine β-synthase
carbon monoxide sensor 3, 4
hydrogen sulfide synthesis 19, 66, 73–75
inhibitor studies in colitis 75, 76
Cystathionine γ-lyase
hydrogen sulfide synthesis 19, 66, 73, 74
inhibitor studies in colitis 75, 76

Diverticulosis, methane production 12
Duodenum
bicarbonate stimulation by hydrogen sulfide 81–89
gas diffusion 7, 8
propargylglycine effects
acid-induced bicarbonate secretion 83–85
acid-induced damage 96
Dyspepsia, *see* Functional dyspepsia

Esophagus, gas diffusion 7
Ethanol, skin gas studies 130
Ethylene, skin gas studies 130

Forest therapy, *see* Phytoncide
Functional dyspepsia, gastric bubbles 10

Gastric emptying, carbon-13 breath test for evaluation 105, 115, 116
Gastrocecal transit time, carbon-13 breath test for evaluation 106
Guanylate cyclase
carbon monoxide binding 3, 36
nitric oxide binding 3, 18, 66

Helicobacter pylori
carbon-13 breath test 104, 105, 114
hydrogen sulfide in pathophysiology 69
Helium, properties and therapeutic application 20, 21
Heme oxygenase
gastrointestinal tract distribution and regulation 37
liver distribution 1
macrophage expression 2
Hepatic stellate cell, carbon monoxide effects 2, 3
High-altitude pulmonary edema, nitric oxide therapy 50
Hydrogen
acarbose effects 97
delivery for therapy
eye drops 96
hemodialysis 97
inhalation 93, 94
injection of fluid 96
oral intake of enriched water 94–97
intestine
inflammatory response 76–78
obstruction findings in breath 11, 12
utilization 9
physiology 92
properties and therapeutic application 20, 92
protective mechanism 93
skin gas studies 130
therapeutic prospects 97, 98
Hydrogen sulfide
bicarbonate stimulation in duodenum 81–89
gastrointestinal function 67, 68
Helicobacter pylori pathophysiology 69
inflammatory response 68, 69, 74, 76–78
leukocyte and neutrophil modulation 76–78
properties and therapeutic application 19, 73
synthesis 19, 66, 73–75, 82
synthesis inhibitor studies in colitis 75, 76
therapeutic potential 69, 71, 78, 79

Hypoxic pulmonary vasoconstriction, nitric oxide response 47, 48

Inflammatory bowel disease
carbon monoxide therapy 38–40
pathogenesis 38
Intestine
gas metabolism 8, 9, 13
obstruction findings 10, 11
Irritable bowel syndrome, intestinal gas 9, 10

Ketone bodies
acetone
expiration correlation with ventilation rate and fat oxidation rate of exercise 120–124
skin gas studies 128, 129
synthesis 119, 120

Left ventricular assist device, nitric oxide adjunct therapy 50
Liver
carbon-13 breath test for function evaluation 105
carbon monoxide effects
sinusoid relaxation mechanism 3
stellate cell 2, 3
nitric oxide therapy in transplantation 52

Metabolic syndrome, hydrogen therapy 96, 97
Methane
diverticulosis production 12
skin gas studies 130

Nitric oxide
acute lung injury treatment 49, 62
acute respiratory distress syndrome treatment 49, 62
administration of therapy 57, 58
bronchopulmonary dysplasia prevention in preterm infants with respiratory distress syndrome 60, 61
cardiac surgery use 50, 62
chronic obstructive pulmonary disease treatment 49, 50
congestive heart failure treatment 50
functional overview 43, 44
guanylate cyclase binding 3, 18, 66
high-altitude pulmonary edema treatment 50
hypoxic pulmonary vasoconstriction response 47, 48
left ventricular assist device adjunct therapy 50
liver transplantation use 52
lung effects 57
properties and therapeutic application 18, 56
pulmonary fibrosis treatment 49, 50
pulmonary hypertension
cardiac surgery association and therapy 61, 62
dysfunction 44, 45
inhalation therapy 46, 47, 62, 63
neonatal therapy 49, 58–60
pulmonary artery pressure response 46, 47
recovery effects of therapy 51, 52
vascular changes and response to therapy 50, 51
skin gas studies 130, 131
synthesis 43, 66
N-methyl-D-aspartate, receptor inhibition by xenon 20

Ozone, properties and therapeutic application 21

Pancreatitis, hydrogen therapy 96
Phytoncide
antioxidant activity 135–137
aroma therapy 139, 140
deodorizer activity 138, 139
forest therapy effect 134
prospects for use 142
soap antibacterial activity 137, 138
solution preparation 135
stress studies in rats
restriction stress 140
water immersion 140, 141
Propargylglycine, effects on duodenum
acid-induced bicarbonate secretion 83–85
acid-induced damage 96
Pulmonary fibrosis, nitric oxide therapy 49, 50
Pulmonary hypertension, nitric oxide
cardiac surgery association and therapy 61, 62
dysfunction 44, 45
inhalation therapy 46, 47, 62, 63
neonatal therapy 49, 58–60
pulmonary artery pressure response 46, 47

recovery effects of therapy 51, 52
vascular changes and response to therapy 50, 51

Sinusoid, carbon monoxide relaxation mechanism 3
Skin gas
acetone studies 128, 129
ammonia studies 129
collection 127, 128
detection systems 128
ethanol studies 130
ethylene studies 130
hydrogen studies 130
methane studies 130
nitric oxide studies 130, 131
types and diseases 124–126
Stomach
gas diffusion 7, 8, 13
gastric emptying evaluation with carbon-13 breath test 105, 115, 116

Therapeutic medical gas, *see also* specific gasses
advantages 16
applications
carbon monoxide 18, 19
helium 20, 21
hydrogen 20
hydrogen sulfide 19
nitric oxide 18
overview 17, 18
ozone 21
xenon 20
definition 15
handling and delivery 17
prospects 21, 22
types and properties 16
Toll-like receptors, carbon monoxide inhibition 19

Ulcerative colitis, *see* Inflammatory bowel disease

Xenon
N-methyl-D-aspartate receptor inhibition 20
properties and therapeutic application 20